Building the Next Ark

Building the next ark
How NGOs work
to protect biodiversity

MICHAEL M. GUNTER, JR.

DARTMOUTH COLLEGE PRESS

HANOVER, NEW HAMSPHIRE

PUBLISHED BY

University Press of New England

HANOVER AND LONDON

Dartmouth College Press

Published by University Press of New England,
One Court Street, Lebanon, NH 03766
www.upne.com

Printed in the United States of America

5 4 3 2 1

Designed by Angela Foote

Library of Congress Cataloging-in-Publication Data
Gunter, Michael M., 1969–
    Building the next ark : how NGOs work to protect biodiversity /
    Michael M. Gunter, Jr.
        p.   cm.
    Includes bibliographical references and index.
    ISBN 1-58465-383-3 (cloth : alk. paper)
        1. Non-governmental organizations. 2. Environmental protection.
    3. Biological diversity. 4. Economic security.   I. Title.
    JZ4841.G86 2004
    333.95'16—dc22   2004013580

This work is dedicated to
Ansleigh Morgan Gunter
and her fellow children worldwide,
both in body and spirit.

We have all been children at one time or another.

May we continue to see the world through your eyes.

Contents

Figures and Tables

Preface

I remember fondly a childhood filled with exploring the outdoors. Late summer days would often be spent navigating a neighborhood stream with miniature wooden canoes, and finding the requisite stones to sink them. As summer turned to autumn, the days grew shorter but would bring refreshingly cool fall breezes and breathtakingly brilliant fall foliage. Winters soon followed with their own seasonal twist as the landscape would be transformed periodically into a magically white wonderland—including the freezing of that local stream as it flowed over a secret cave to create magical crystals for admiration or mighty ice swords to detach for combat. In time, spring and summer months would then fill the days with an assortment of hiking, camping, and swimming before the entire cycle repeated itself.

I suppose it was at this early age that an appreciation for the environment was instilled in me. To my knowledge, we didn't have any highly unusual mega-fauna or endangered flora in Middle Tennessee, but we surely had plenty of waterfalls and gorges and wind-weathered cliffs to explore. And as I know better now, there were likely large numbers of species that escaped my eye back then, for life on Earth comes in many colors, many shapes, and many sizes. These experiences resonate all the more with me now as it becomes painfully clear that these cherished experiences are threatened the world over.

On the one hand, this is a very trivial connection, for biodiversity has much, much greater significance for our environment and the health of the people that are, by definition, part of that environment. But on the other hand, as my research demonstrates, this "trivial" connection also carries a high degree of significance. Perhaps the most challenging obstacle to protecting biodiversity is to get people to care. When considerable portions of this biodiversity exist in far-away places, this can be an imposing task. The trick, then, is to draw a local connection, to connect domestic and international interests. People must be shown why they should care about an issue based far away. Those who seek to protect biodiversity must convince people of the reasons why they should care. They must demonstrate that this biodiversity is close figuratively if not literally.

Perhaps the most universal link in accomplishing this task rests with the youngest generation, our children. This is the one constituency that virtually every society can agree deserves attention, a fact not lost on the environmental community and their growing number of youth initiatives. But I believe we as a society gain not only from teaching our children about the environment. We can learn about the environment from our children as well. Children see the world differently than adults. They are physically smaller, so their vision is from a decidedly lower perspective. Because children are younger than adults, they ususaly have less experience than their elders. We tend to think of these characteristics as negatives, and there are many cases where this is true. Experience is to be cherished. We learn from it. We improve ourselves. It can also be a tremendous benefit to view your surroundings from higher ground. To see from a "taller" perspective enables an adult to consider more of the their surroundings whereas a child's perspective is much more limited.

But one could easily twist this superficial assessment condemning childhood perspective to one that hails it. Sometimes it is better to have a "smaller," more limited focus, to keep yourself grounded in the more immediate. Experience can also harden individuals so that the everyday wonders of life are taken for granted. My point here is not to suggest that ignorance is bliss. In fact, it is entirely the opposite. We can learn from our childhood and our own children, every day. As any newly anointed parent will boast, there are so many ways in which this is true. The single most important aspect to highlight here is that children see the world through young eyes. They touch their first blade of grass with awe. They twirl sticks for the first time with pure joy. They giggle and jiggle with delight when a gecko puffs out its cheeks in the hot Florida afternoon sun. They take their first dip underwater with a mixture of fear, uncertainty, and bravery. They cherish the new, the undiscovered. One could argue, then, that there is a decided advantage to being closer to the ground, to seeing the world with a "smaller" and "less experienced" perspective. Sometimes we adults get lost in past experiences or distracted from the most immediate surroundings by trying to see too far. Sometimes a "shorter" perspective or "less" experience is more useful.

The analysis in this book is not about children, of course, but it is about the need to retain awe and wonder toward our environment. Biological diversity is threatened worldwide today, in numbers so great that we are faced with the sixth major extinction spasm in history. Half the world's tropical rainforests are already destroyed—particularly troubling as this is home to the vast majority of species diversity in the world today. Some fifty species a day are destroyed by deforestation in these areas alone. Add those lost to overall habitat loss, invasive species, or general pollution, and numbers of truly frightening proportions emerge. Yet too often this goes unnoticed. And we raise these risks all the more. Fortunately, though, an increasing number of nongovernmental organizations (NGOs) hope to reverse this neglect. NGOs hope to awaken a slumbering public to the tragic biodiversity loss that is unfolding today. Indeed, NGOs are the most appropriate medium for protecting biodiversity because of their unique ability to see both "small" and "big." They operate at both the micro and macro levels, working on the ground in local villages as well as participating in international negotiations. It is in this respect that this class of actors, NGOs, is very childlike in a constructive sense. We have all been children at one point in time. Let us not forget what that sense of awe and wonder felt like, for our own sake, and theirs.

Winter Park, Florida *M.M.G., Jr.*

Acknowledgments

Not surprisingly, my work on this book holds several characteristics in common with the issue that it examines. To study the effectiveness of NGO biodiversity protection strategies, one must incorporate a multitude of perspectives, from the growing number and relevance of environmentally oriented actors, both governmental and non-governmental alike, to an expanding academic literature in both environmental studies and international relations. Reflecting the traits of both ecological and political interdependence, then, this work was only possible with the assistance of an array of individuals in each of these arenas.

First and foremost, let me extend my appreciation to those in the environmental community who dedicate their lives to biodiversity protection and shaping the policies that seek to promote it. While there are too many to name in this short space, and some must remain anonymous for their candid insights, I would be remiss if I failed to call attention to a few names that opened up the doors of their respective NGOs, foundations, and governmental bodies with welcome arms. Thanks to David Guggenheim, Bob Irvin, and Carol Gardner at the Ocean Conservancy; Rod Mast, Haroldo Castro, Keith Alger, Cyril Kormos, Christian Heltne, and Sterling Zumbrunn of Conservation International; Hans Verolme of Biodiversity Action Network; Fabienne Fon Sing and Sam Johnston of the Secretariat of the Convention on Biological Diversity; Carroll Muffett and Rina Rodriguez at Defenders of Wildlife; Marie Studer, Blue Magruder, and Jeff Venier at Earthwatch Institute; Tom Turner and Shana Glickfield at Earthjustice Legal Defense Fund; Reshma Prakash at *The Earth Times*; Michael Bean, Bruce Rich, and Korinna Horta at Environmental Defense; Peter Spain at AED's GreenCOM; Sanjayan Muttlingyam, Mary McClellan, and Melissa Ryan at The Nature Conservancy; Steve Mills, Dan Seligman, Ellen Bryne at Sierra Club; Seema Paul and Nicholas Lapham at United

Nations Foundation; Franklin Moore and Elise Storck at USAID; Scott Hajost at IUCN; Melissa Boness, Bret Bergst, and Nels Johnson of World Resources Instiutute; Brooks Yeager, Estraleta Fitzhugh, Ruth Franklin, and Wes Wettengel of World Wildlife Fund; and Mick Seidl of the Moore Foundation. As different as these individuals are, much like the organizations they represent, they do all share at least one common trait—a commitment to ensuring the best possible future for succeeding generations. Thanks also to independent artists Becky Bornhorst and Kirk Anderson.

From the academic world, my mentors at the University of Kentucky also deserve many thanks for their patience and prodding over earlier stages of this manuscript. In particular, I would like to thank Karen Mingst and Ernie Yanarella, for having faith in me during my doctoral student days. Karen Mingst was a model committee chair, and made earlier versions of this work stronger through her theoretical guidance and tireless support from the very genesis of this idea and what must have seemed like endless drafts along the various writing stages. Her feedback on various chapters even as she was out of the country on sabbatical for her Distinguished University Professor award is much appreciated. Special thanks also to Ernie Yanarella, who has been both indispensable and selfless in devoting his time and advice as I embark on my own career in academia. His encouragement to speak in my own voice, in particular, made it possible for this work to reach a larger audience today. Also at the University of Kentucky, Phil Roeder, Horace Bartilow, Ramona Rush, and Edward Jennings deserve thanks for their assistance in navigating various bumps along the way. Latter stages of this project, moreover, were possible thanks to several individuals at Rollins College and University Press of New England. Here at Rollins College, Dean Roger Casey generously extended financial assistance to complete a number of follow-up personal interviews that enhanced this work immeasurably. And thanks to Austa Weaver and my colleagues in the department of political science, particularly Tom Lairson, for his guidance and support in my time at Rollins. Thanks also to those who reviewed this manuscript and the editorial, production, and marketing staff of University Press of New England, namely senior editor Phyllis Deutsch, who helped shape this work from mere proposal stage to the completed work before you now. This book is truly a better work due in no small part to her.

Thanks finally to my family as well as extended family for their emotional support and advice in the production of this book. My mother and father, Judy and Michael Gunter, deserve special mention. In so many respects, this work is an extended dedication to them. They provided perspective and motivation in completing this project, probably in more ways than they will ever know. And I truly have had the benefit of seeing how books develop over time thanks to the accomplished author for whom I am named, my father. If this work achieves some small measure of greatness, it is only because I have stood on the shoulders of those who came before me. (Here I paraphrase Sir Isaac Newton's famous reference to his dependency on Galileo's and Kepler's work in physics and astronomy.)

My father has been and always will be, my role model. Without his candid criticism and warm encouragement, this book definitely would not be before you today. My mother, similarly, taught me both compassion and perserverance, even in the face of the toughest of odds. She is my hero. Most of all, let me extend my deepest thanks to my wife, Linda, and our daughter Ansleigh. Linda understood how important this work was to me and supported me unconditionally throughout the entire process. She is a trusted confidant, who offers keen insights and an intoxicating laugh on a daily basis. She is my best friend and true love. It is my own selfish hope that, in its own small way, this book will help make the world a better place in which we live—together.

Building the Next Ark

Introduction:
Why a new ark is needed

A NEW ARK

A new ark is needed. Ominous storm clouds have gathered to threaten humanity's most basic resource of all—the diversity of life on Earth. With a flood of potentially biblical proportions confronting us today, with a United Nations estimate of fifty thousand species extinctions a year, policymakers are under more pressure than ever to develop viable options to preserve the diversity of species that exists on Earth. We need these species for any number of reasons, reasons that run the gambit from moral values to economic values and from ecosystem services to human health. The blood of a horseshoe crab helps in the diagnosis of meningitis. Bee venom aids in treating arthritis.[1] And there is so much more we do not yet understand. But playing Noah is no easier the second time around. Protecting biodiversity is widely recognized as problematic for both ecological and political reasons. Ecological threats often stem directly from political hurdles, and political differences, in turn, increasingly find their root cause in ecological questions and controversies.[2] The complex and interdependent nature of biodiversity protection demands more than a cursory discussion of political and ecological dimensions as distinct entities. A truly complete understanding of this issue requires a thorough consideration of the actual linkages between politics and ecology as well. Nowhere is this more evident than in the transnational realm, where environmental issues often transcend state boundaries.[3] Indeed, the emerging phenomenon of global environmental threats rightly calls into question the extent to which individual state governments, if they insist on acting alone, can properly deal with this new order of policy problems.

A burgeoning literature suggests that nongovernmental organizations (NGOs) may serve as the missing link in this equation, that NGOs provide valuable state assistance by fostering a series of fundamental linkages required for effective biodiversity protection at the transnational level. These linkages are the connections between domestic and international, ecological and economic, and short- and long-term considerations. While their significance at the macro level is widely recognized within environmental studies, these linkages have received scant attention at the micro level— particularly in political circles. That is, while scholars have made notable contributions in terms of outlining the three essential functions that NGOs provide, we still know very little about how the day-to-day operations of NGOs translate into their ultimate objectives.[4] The dominant question as to how NGOs most effectively enhance biodiversity protection remains. How do NGOs help build the next ark?

TELLING THE STORY

That question in turn raises several follow-up questions. What strategies work best? What strategies do not work as well as NGOs would hope? And how can NGOs adapt accordingly? Positing that the three fundamental linkages need to be made—namely links between domestic and international, ecological and economic, and short- and long-term considerations—this book examines how both specific mainstream and participatory strategies attempt to enhance transnational biodiversity protection. It centers on the issue of strategic effectiveness, and offers prescriptions for how NGOs can improve their protection efforts. Yet, not only is this book an analysis of how and why NGOs apply mainstream and participatory strategies. It also suggests how NGOs may adjust their own organizational characteristics to ensure that they support those strategies rather than constrain them. This book looks at how NGOs work within the system. It looks at how NGOs work with people. And it looks at how NGOs must work on themselves.

This analysis tells a story, then, of how NGOs themselves fit within the larger environmental activism community. It details seven semi-discrete strategies employed by eleven different environmental NGOs, finding that the most impressive results come when mainstream and participatory strategies are paired with one another. One NGO need not make these pairings between mainstream and participatory strategies alone. An NGO is better off, in fact, concentrating on a specific strategic niche that may or may not involve both a participatory and mainstream strategy component— as long as another group is providing the complementary strategy. Thus, an organization's willingness to form partnerships without sacrificing concerted strategic concentration is critical in determining strategic effectiveness. This allows NGOs to concentrate on their specific niche yet still enjoy

Table 1. Environmental NGOs Examined

	Headquarters	Date founded
Biodiversity Action Network	Washington, D.C. (defunct at end of 2000)	1992
Conservation International	Washington, D.C.	1987
Defenders of Wildlife	Washington, D.C.	1947
Earthjustice Legal Defense Fund[1]	Oakland, Calif.	1971 (1997)
Earthwatch Institute	Maynard, Mass. (Boston area)	1971
Environmental Defense[2]	New York, N.Y.	1967 (2000)
The Nature Conservancy	Arlington, Va. (D.C. area)	1951
The Ocean Conservancy[3]	Washington, D.C.	1972 (2001)
Sierra Club	San Francisco, Calif.	1892
World Resources Institute	Washington, D.C.	1982
World Wildlife Fund	Washington, D.C.	1961

[1]Earthjustice Legal Defense Fund changed its name from Sierra Club Legal Defense Fund in 1997 to reflect, among other things, that the Sierra Club was not its sole client. A separate entity since its founding, Earthjustice still retains a close relationship with the Sierra Club but believes the new name to be more inclusive. Chapter 4 will describe this adjustment in further detail.

[2]Environmental Defense Fund changed its name and logo in January 2000 to avoid confusion among some who equated the term "fund" with a government agency. Its name was shortened to Environmental Defense along with the subtitle "finding the ways that work." Environmental Defense also adopted the "e" from its name as a symbol of the Earth and other recurring themes in its work—environment, economy, equity, and empowerment. Many in the environmental community, though, still refer to the group by its popular acronym, EDF. This underscores the trade-off to making such name changes as some degree of past identity is often compromised.

[3]The Center for Marine Conservation changed its name to the Ocean Conservancy in mid-2001 after discovering in opinion polls that some thought the word "center" in CMC indicated it was a governmental or academic entity. A number of others were confused about the term "marine," mistaking CMC as a military organization. The name change is not entirely immune to terminology baggage, though, as some now assume Ocean Cosnervancy is part of The Nature Conservancy.

the synergies created by combining mainstream and participatory strategies together. That said, one should also note that a mix of NGOs operates within the environmental community. Those above represent diversity in terms of size and strategic emphasis, among other characteristics.[5] Table 1 highlights the eleven environmental NGOs examined.[6]

In fairness, it should also be noted what this book does not do. This study does not detail which species are being lost. Neither does it include an exhaustive list of environmental NGOs working in the biodiversity protection arena—even those simply based in the United States. In fact, several notable biodiversity-focused NGOs such as the New York City–based Wildlife Conservation Society and the legal think tank the Center for International Environmental Law in Washington, D.C., and Geneva, Switzerland, are left out. The analysis, moreover, does not pretend to detail the entire history and complete array of activities of the eleven groups this study does utilize. Other exemplary works exist in this realm. Just a few of note are Michael Cohen's impressive history of the Sierra Club, Tom Turner's powerful description of Earthjustice, William Weeks' reflection on years of experience with The Nature Conservancy, and Paul Wapner's fascinating academic comparison of Greenpeace, Friends of the Earth, and World Wide Fund for Nature (WWF-International).[7]

One final disclaimer of note is that this work looks at groups that operate globally—but from a headquarters in the Northern Hemisphere, specifically the United States.[8] This is done for two reasons. For one, the vast majority of funding for biodiversity protection comes from the developed states of the North. For another, the United States is still the preeminent global power in any number of measures of that term. NGOs understand that to really make a difference they must have a foothold in the United States, for both financial and political reasons. As Seema Paul, senior program officer for biodiversity at United Nations Foundation, contends, if NGOs are not based in the United States, they operate from a distinct disadvantage.[9] This emphasis, however, should not be construed as denying the critical importance of Southern Hemisphere NGOs. Indeed, a central argument in this book is that biodiversity protection is only possible when domestic and international connections are made. This highlights the role of partnerships in the NGO community, including partnerships between Southern and Northern NGOs. Some Southern NGOs, in fact, actively seek out Northern allies to apply international pressure on their domestic governments. Know as the "boomerang strategy," this can be particularly effective as long as sufficient democratic structures are in place.[10] Admittedly, one must also be cognizant of dangerous political misperceptions here. Any emphasis on the North when it comes to these partnerships leaves one open to charges of neo-imperialism, or as it is known in environmental circles, green imperialism. Northern NGOs struggle with this themselves daily. And this analysis in no way means to suggest that Northern NGOs hold all the answers to species protection, while Southern NGOs need only listen to them. Nothing could be further from the truth. This must be a true partnership where equality and diversity are celebrated together.

What this analysis does add, then, is a more complete picture of the requisite political paths for effective protection efforts by examining eleven environmental NGOs that emphasize transnational biodiversity protection

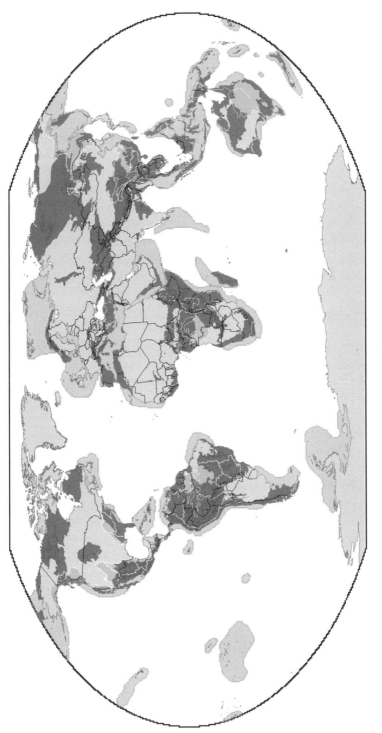

Figure I.1. WWF's two hundred global eco-regions. SOURCE: WWF Conservation Science program.

in their strategic vision and specific policy initiatives. While the existing theoretical research provides much support for NGOs as the proper medium to pursue species protection internationally, it offers little insight as to how this process develops politically. Yet that, in and of itself, is valuable information as noted in my opening contention that ecological and political considerations are fundamentally linked when it comes to species diversity protection efforts. Figure I.1 visually underscores this fact as endangered species and their habitat fail to fall neatly within the confines of state political boundaries. Of course, the underlying thesis in this book is that this overlap is "shared," in turn, with even those that are not physically housing these species. Global environmental change and pollution may originate from a nonadjacent region or even a completely different ecosystem on the other side of the world, as is the case with climate change, acid rain, and ozone depletion in addition to the biodiversity loss focus of this study.

THE BATTLE AHEAD

WWF traditionally identifies two hundred global eco-regions, the most biologically distinct terrestrial, marine, and freshwater regions according to their scientists. Distinctiveness is measured according to total number of species, number of species found nowhere else (referred to as "endemism"), and presence of unusual evolutionary or ecological phenomenon. The approach reflected in their map differs from the one taken by the group Conservation International (CI), whose members believe the best strategy is to concentrate upon a relatively small number of areas with high species diversity. Their "hot spots" strategy currently pinpoints twenty-five biodiversity locations that cover a mere 1.4 percent of the Earth's land surface—but land that is home to more than 60 percent of the global terrestrial species. This decision has enormous political repercussions, though, as it leaves out large tracts, areas that house WWF's Global 200 program.

Politics permeates this issue. As we shall see in the ensuing pages, human health, economic development, and cooperative state relations—not to mention rudimentary ethical issues—are all at stake when considering the vital resource of biodiversity. Ironically, though, even as we improve our understanding of the value that species diversity holds, we have much to learn about the very number of such species with which we are dealing. Only a small percentage of the species alive today are classified scientifically. Some 1.7 million species are known, with another three to thirty million different species remaining undocumented according to even the most conservative estimates. More liberal estimates place current species diversity in the neighborhood of 100 million different species.

Juxtapose these numbers with the aforementioned estimate that fifty thousand species are lost every year and a frightening picture emerges.[11] That is why biodiversity protection is so important. In many instances, the species we lose are not even known, let alone fully understood in terms of their position and functions in their respective ecosystems. In simple terms, we do not even know what we are losing. One can take this argument a step further and contend that many such losses will only be understood once it is too late, when the option to conserve and protect is no longer available. This danger highlights the threat of irreversibility and lends further credence to the emerging soft law concept known as the "precautionary principle," a term best articulated in Principle 15 of the 1992 Rio Declaration at the United Nations Conference on Environment and Development. It states:

> In order to protect the environment, the precautionary approach shall be widely applied by States according to their capabilities. Where there are threats of serious or irreversible damage, lack of full scientific certainty shall not be used as a reason for postponing cost-effective measures to prevent environmental degradation.[12]

The following chapters discuss transnational biodiversity protection efforts within this context, specifically the promise that particular NGO strategies hold in curbing this extinction crisis. NGOs are uniquely suited to address three main obstacles to effective biodiversity protection. These hurdles are a fundamental disagreement over the conceptualization of power and self-interest, ambiguity inherent within the term "sustainable development," and lack of a global ecological consciousness. The first of these obstacles centers upon the omnipresent international relations debate over the dominant realist paradigm, namely disagreement over how constricting power and self-interest are when it comes to cooperation at the global level. This is particularly true in that it challenges the theory of relative gains posited by those such as political scientist Joseph Greico.[13] As the fundamental theoretical conflict in transnational relations, liberal versus realist interpretations of power and self-interest attract a wellspring of academic activity. In particular, with the changing status of power and governance at the close of the twentieth century, neo-liberal approaches incorporating international organization, regimes, and non-state sources of power continue to gain adherents within the discipline of international relations. But the bulwark of state sovereignty, namely the continued primacy of realist self-interest and power in transnational relations, also continues to impede the progress of both established and emerging liberal cooperation regimes, that is, a transnational biodiversity protection regime, and the power sharing institutions that foster it, such as the 1992 Convention on Biological Diversity (CBD).[14]

Confounding biodiversity protection efforts further is a second obstacle, the highly controversial and deeply ingrained value judgments that surround assessment of the impact that species protection has on economic efficiency. Economic development too often becomes thoughtless, unchecked development that fosters a cancerous paradigm where growth suffocates sustainability. In this case, of course, development is not even the appropriate term. What is really meant is economic growth, growth that clashes with environmental regulation in the definition over what is a beneficial or even an acceptable level of development. A vast and divergent sustainable development literature exists in this regard. Those such as former World Bank economist Herman Daly outline perhaps the most widely accepted definition. Daly describes sustainable development as development that utilizes the interest of the Earth's natural resources without encroaching upon its capital.[15] While the beautiful simplicity of this analogy continues to draw converts rhetorically, the exact steps for implementing such a policy recommendation remain hotly contested. No consensus exists on precisely how this theory translates into action. Much of this centers on what Herman Daly and co-author John Cobb identify as the "fallacy of misplaced concreteness in economics." This economist and theologian, respectively, argue that economic theory emphasizes money and markets rather than actual physical goods, with the result being a model that encourages unsustainable development.[16]

Lastly, but perhaps most importantly, achieving an elusive yet essential ecological consciousness that transcends the national political boundaries of today is critical to protecting biodiversity. This acknowledges that the subject of this study is truly a non-parochial one. We all have a stake in preserving the diversity of life on Earth. Yet, precisely because of traditional state sovereignty and the neo-classical liberal economic paradigms outlined above, cooperative actions for the common good are difficult to achieve. It is in overcoming this third and final obstacle that NGOs perform perhaps their most noteworthy function by enhancing communication efforts. More specifically, NGOs create and maintain a global civil society that fosters the requisite level of ecological consciousness to effectively protect transnational biodiversity. Global ecological consciousness here simply refers to an ecological understanding of the relationships between local and global actions, an understanding that transcends national boundaries. According to a global ecological consciousness, the world is recognized as an organism where each of its bodily parts is dependent on the others. Components are not discrete, nor are they disposable.

Despite its crucial status, though, scholarship on global ecological consciousness has failed to receive anywhere near the same level of attention as our first two obstacles. This neglect is a notable shortcoming, as, without global ecological consciousness, questions of political participation and financial constraints cannot be addressed in an equitable manner, one that recognizes North-South discrepancies in protection costs and the local-global tensions in implementing biodiversity protection programs. Long-term resolution of transnational environmental issues like biodiversity protection requires precisely this equitable approach. This is particularly true when one considers the clear discrepancies in the amount of biodiversity between Northern and Southern hemisphere states. The first world, the developed states with the richest economies of the world, reside largely in the North. The poorer third world or developing states fall mostly into the Southern Hemisphere geographically—*and* hold most of the global biodiversity.

THEORETICAL CONTEXT

A wealth of academic literature positions NGOs as the best, perhaps only, medium suited to bridge this North-South divide. NGOs are the most effective vehicles to negotiate the treacherous terrain ahead for a number of reasons. NGOs can broker state sovereignty with global community concerns. NGOs can open up the requisite participation channels. NGOs can sow the seeds for an ecological consciousness that will foster future transnational environmental agreements with real enforcement mechanisms. This is a story about this process. It is a story about NGO effectiveness in achieving strategic niches and balancing partnerships with other organizations. Only then will NGOs be able to leap the hurdles of conflicting self-interest and power, ambiguous sustainable development, and global ecological unconsciousness.[17]

This turn to NGOs is a relatively recent phenomenon. While non-state actors such as NGOs have been important players on the international stage for decades, their role has changed dramatically in the last twelve to fifteen years. This is true for a number of reasons. Several specific events of note are the 1989 end of the Cold War and subsequent 1991 dissolution of the Soviet Union, internal momentum from a buildup to the 1992 United Nations Conference on Environment and Development (UNCED or Earth Summit), and the contentious 1993 debates regarding labor and environmental side agreements to the 1994 North American Free Trade Agreement where two disparate groups were brought together.[18] These combine in the mid-1990s with an increased understanding of environmental issues as well as better access avenues to data on those issues, namely in the form of the Internet. NGOs, now more than ever, are crucial components to understanding world politics.

Adopting this NGO perspective also supports the long requested call by George Washington University's James Rosenau for linkages between the domestic and international dimensions of political science.[19] And it outlines the necessary components for future foreign policy decision making. While the series of connections that exist today complicate the decision-making process and at times directly challenge state sovereignty, states do continue to operate, even manipulate, the complex interdependence first identified by Duke University's Robert Keohane and Harvard University's Joseph Nye.[20] It is precisely this relationship that the following analysis emphasizes in focusing upon NGO influence with regard to the state, although this proposal also recognizes the critical nature of relationships above and below the state. Echoing past researcher sentiments from James Rosenau to UC-Santa Barbara's Oran Young, this book argues that the challenge of biodiversity protection encompasses a new generation of problems within the transnational arena, one that has yet to find a new generation of solutions. That is why a new ark is needed.

The transnational and interdependent nature of biodiversity protection

ECOLOGICAL SETTING

Biodiversity Defined

No issue illustrates the need for more than a simple, state-centric perspective better than the threat of biodiversity loss, what many scholars identify as the single most important policy hurdle of the twenty-first century.[1] First coined in 1986 by Walter Rosen, administrative officer at the National Academy of Sciences, the term "biodiversity" moved quickly from scientific circles to the popular lexicon by the late 1980s. Numerous authors, from Harvard geologist Stephen Jay Gould and Stanford biologist Paul Ehrlich to science journalist Yvonne Baskin and renown Yale social ecologist Stephen Kellert to Niles Eldridge, curator in the Department of Invertebrates at the American Museum of Natural History, have made notable contributions in the years since. But it is Pulitzer Prize winner E. O. Wilson whose name is most synonymous with the term today. After 1988, Edward O. Wilson, with his edited work, *Biodiversity*, became the leading authority on the concept. According to Wilson, "Biodiversity . . . is the total hereditary variation in life forms, across all levels of biological organization, from genes and chromosomes within individual species to the array of species themselves and finally, at the highest level, the living communities of ecosystems such as forests and lakes."[2] One must be careful to note that within this context "biodiversity" by no means implies static balance. No such equilibrium exists in nature—contrary to popular interpretations. The only constant is change. Streams dry up. Rivers flood. Volcanoes erupt, then smolder or lie dormant. Mountains erode. Lakes evaporate and silt. Climate warms and cools. Diversity, by definition, changes.

Natural disturbances are often responsible for such change. Some changes are subtle. Others are not. Fires, storms, and hurricanes, for instance, all assist in dramatic forest renewal and natural replacement. With-

Figure 1.1. Yellowstone's scarred landscape? Michael M. Gunter, Jr., Yellowstone National Park, July 2001.

out such processes, forest ecosystems undoubtedly would suffer. The rich soils of the forest floor depend in large part on the outbreak of small forest fires, for burning and charring replenish much-needed nutrients. Old-growth stands are mature enough to withstand periodic forest fires of limited intensity; yet, when too much underbrush is allowed to accumulate within a forest, when forest flames do not ignite for long periods of time, even old-growth stands are susceptible to outright destruction. They simply are not equipped to withstand the intensity of the fires that inevitably will alight.

The Yellowstone fires in the summer of 1988 are a case in point. Years after this "national tragedy," in fact, hillsides like the one shown in figure 1.1 still show the effects of that devastating fire. Following decades of fire suppression by the National Park Service, new policies allowed for natural fires to burn out on their own. But the summer of 1988 was the driest in the 112-year written record of the park, and a number of natural, lightning-caused fires failed to burn out. The United States government enlisted 25,000 people, including army and marine units, and spent $120 million fighting them, making it the largest fire-fighting effort in U.S. history. Only September snows finally stopped the fires. Interestingly, despite what seemed like a disastrous burning in 36 percent of the park, many species benefited immensely in the following years. Perhaps beauty truly is in the eye of the beholder. What appears to the typical backpacker to be little more than scorched scars across the landscape just as easily can be labeled beauty marks by ecologists studying the region today. Lodgepole pine, in fact, can only reseed itself when fires burn hot enough to force its cones to open.[3]

It is this level of disturbance that a variety of academic disciplines, including transnational relations, now address. As modern industrialization has made all too clear, inanimate objects are not alone in their influence upon the environment. Humanity also inflicts a substantial degree of change. Throughout the world, human development threatens the biodiversity of our planet. Some human "fires" are natural, of course. They are unavoidable, as humanity is part of the ecological system. And some are needed, serving as integral components in a complex ecological equation. When we substitute and act as the tornado, hurricane, or firestorm, this is *not* an evil to be avoided at all costs. Change and death, even when purposefully instigated by humanity, are naturally expected and accepted—up to certain sustainable levels. What is of direct concern, however, is the fact that human "tornadoes" and "hurricanes" are occurring with ever increasing frequency and destructive impact as the twenty-first century opens. The scale and frequency of these actions are *not* sustainable. The current level of change, then, is not problematic in that it creates death and destruction. It is a crisis because the *rate* at which death and destruction are occurring makes natural renewal impossible.

A small minority of scholars, however, question the crisis context in which this definition of biodiversity loss is presented. They contend that the loss of diversity currently underway is only natural, that human responsibility is minimal. Pointing to past examples of evolutionary shifts, this school of thought notes that current biological changes represent a mere blip in geological time. Authors such as journalist Charles Mann and economist Mark Plummer argue that a more accurate portrait of the process underway must take into account this key distinction instead of the more common generation-specific approach to policymaking.[4] This is an

extremely effective point with sound scientific reasoning. The only drawback is that such approaches ignore the magnified rate of habitat loss today,
do not address questions of irreversibility, and display an explicit disregard
for the aforementioned precautionary principle—the emerging international soft law principle that humanity should take every precaution to
avoid environmental damages that may prove to be uncorrectable at some
point in the future. This also calls attention to a critical assumption of
those within the "wait and see" camp. According to their position, environmental regulations are only necessary once damage is proven, which
implicitly suggests that either humanity can adequately compensate for
those losses incurred, or that the species that are currently being lost are
marginal or unnecessary components in the web of life.

An examination of several environmental economics works outlines this
rationale. Scholars such as 1976 Nobel economics prize winner Milton
Friedman, Hudson Institute founder Herman Kahn, and conservative
business economics champion Julian Simon assert that any energy supply
problems can be corrected because resources are not limited—and even if
they were, such natural limits could be surmounted by boundless human
ingenuity.[5] Building on the works of Francis Bacon and René Descartes,
this paradigm positions technological innovation as the ultimate savior for
any environmental transgression, including species loss.[6] Crisis is but a
challenge to human ingenuity for those in this cornucopia camp. Friedman
even contends that the crises of today are simply crises of confidence, that
they are no different than previous challenges in which humanity has tri-
umphed. Citing the deforestation of British forests, Friedman describes
how the rise of coal to replace wood as an energy source avoided the first
modern energy crisis. Similarly, discovery and application of petroleum sup-
planted coal before its reserves were depleted. The next energy crises, according to Friedman, will be solved in much the same manner.[7] It is within
this vein of thought that Charles Mann and Mark Plummer address biodiversity protection. Operating under the human ingenuity parameters of
Friedman and others, these co-authors describe a species protection program where sacrifices must be made according to worth for humanity.

This logic relies heavily upon cost-benefit analysis, a dangerous route,
as seen in the critique of this practice by Amy Lovins, chief executive officer at Rocky Mountain Institute. Cost-benefit analysis is intrinsically sub-
jective, often neglecting hidden costs (such as remedial and transaction
costs) within stated benefits. Its attempt to break down complex interactions into their individual components also discounts holistic relationships.[8] Adopting past interpretations of an impermeable boundary be-

tween man and nature, this school of thought contends that distinctions can be drawn between the two and exploitation of one can take place without consequence for the other.

Historically, the environment has been considered as something apart from humanity. As Berkeley geographer Clarence Glacken explains, Christianity and capitalism combined to perpetuate this thinking for centuries. Wild areas were considered dangerous, even wicked. Terms such as swamp, jungle, and desert emerged in the lexicon, carrying unsavory connotations. Civilization was civilized precisely because it had tamed its wild surroundings. Gardens and lawns were decidedly more peaceful and less threatening. Implicit, and often times even explicit in this thinking, was the rationale that nature could be conquered by man (sexism intended).[9] And as Leo Marx points out, this compartmentalization can actually decrease efficiency and creativity for the simple reason that people often become pessimistic and inactive when technology instead of them is dictating the future.[10] In the words of Henry David Thoreau, "men become the tools of their tools."[11]

To avoid this danger, the modern environmental movement, with nineteenth-century philosophical origins in the work of conservationist George Perkins Marsh and the transcendentalist Thoreau, emphasizes a holistic approach.[12] Stressing the need for a healthy relationship between humanity and the environment, this philosophy is firmly grounded in the concept of interdependence. Domestically, this foundation was set in 1962 with the publication of Rachel Carson's groundbreaking text, *Silent Spring*. This work galvanized the environmental movement in the United States by focusing on a spring when the birds no longer sang due to the indiscriminant use of highly toxic pesticides. But perhaps most importantly in terms of long-term effects, *Silent Spring* cemented public consciousness regarding the interdependent nature of the environment.

The Importance of Biodiversity Protection

Despite an emerging consensus regarding the interdependent nature of our environment, considerable disagreement remains as to what worth biodiversity really holds, both ecologically and economically. This debate centers on an alarming scientific uncertainty that permeates all work on this issue, beginning with the very number of species we are considering. To date, some 1.7 million known species exist. The great majority of species alive today, though, possibly as much as 90 percent, are not known.[13] Depending on which expert estimate one applies, that translates into anywhere from another three million to one hundred million species upon the Earth at this time. While actual species types are constantly changing, the general number of species has remained constant in the modern human era—until now. Constancy was made possible by speciation, a natural influx that ensures that new species emerge each year to offset expected extinctions. Speciation is a long process, which develops as a species colonizes new regions far from other populations of that species.[14] Environmental pressures then shape a genetically different evolution tied to this reproductive isolation. In very basic terms, speciation is a random process of mutation but also a deliberately efficient act of natural selection that determines which mutations are eliminated and which survive.

Organisms such as the horseshoe crab have survived this process for the past 200 million years. Others such as the dinosaurs were not as fortunate. Then again, *this* particular extinction worked out well for humanity. It "allowed obscure primitive mammals to evolve into [their] dominant role of today."[15] A major concern here is the possibility that history may repeat itself, only this time humanity would find its present favorable position reversed. Such a scenario deserves added consideration given current levels of growth, levels that by virtually all accounts are unsustainable. Continued population increases and virtually exponential human encroachment upon wildlife habitat severely threaten our environmental foundation. Entirely of our own accord, unchecked excesses could trigger an irreversible spiral, landing humanity at the losing end of the evolution game.

In very basic terms, in order to adapt to a changing environment, the raw materials of nature as well as humanity itself, require genetic, species, and ecosystem diversity. This diversity "increases the likelihood that at least some species will survive and give rise to new lineages that will replenish the earth's biodiversity."[16] While scientific evidence remains inconclusive, all preliminary indications suggest that recuperation capabilities will be compromised severely if biodiversity is not protected. As Wilson asserted over twenty years ago, the ramifications would be catastrophic: "The one process . . . that will take millions of years to correct is the loss of genetic and species diversity by the destruction of natural habitats. This is the folly our descendants are least likely to forgive us."[17]

Despite this potential for irreversibility, though, some scholarship continues to question the financial utility of conservation efforts, particularly prohibitively expensive endeavors that limit economic development. This school of thought turns to the aforementioned cost-benefit analysis and the series of worth judgments that it dictates. It is within this context that the sustainable development literature provides valuable insight. The term emerged in the international lexicon with a 1980 publication of the hybrid NGO-state organization known popularly as the World Conservation Union

(IUCN), a serendipitous source considering the role NGOs play in this debate today. By 1987, the Brundtland Commission, led by Norwegian Prime Minister Gro Brundtland, had popularized this sustainability approach with its widely circulated, UN-sponsored text, *Our Common Future*.[18] Essentially, as economist Herman Daly and theologian John Cobb contend, the Brundtland Commission described sustainable development as living off the interest of the Earth without encroaching upon its capital.[19] Scholars such as University of Kentucky's Ernest Yanarella and Richard Levine have further refined this definition to explicitly incorporate "maintenance" of the Earth and its resources.[20] Similarly, Hazel Henderson writes about shaping environmental policy from the perspective that we are borrowing from our children.[21] Following the logic of James Lovelock's *Gaia*, these approaches adopt the perspective that the Earth is a dynamic organism that needs its own resources to continue to flourish—and that human activities must take this into account.[22]

The Brundtland Commission capitalized on this academic activity, utilizing past fears of environmental degradation and questioning the rationale guiding endless economic growth. Donella Meadows, for instance, ignited a veritable explosion of literature in the 1970s regarding the institution of limits for the entire planet, humanity included.[23] Paul and Anne Ehrlich's "rivet hypothesis" adopted this same perspective as the co-authors compared Planet Earth to a plane that one by one was losing its rivets, or, in the case of Spaceship Earth, its species diversity. Individually, loss of these rivets/species would probably have only negligible effect. In time, however, these losses would accumulate to such a point that the plane/Earth would no longer be sustainable. It would crash. William Ophuls, similarly, addresses the carrying capacity of the Earth in outlining three possible scenarios for the future: one rosy, one realistic, and one apocalyptic. Building upon the work of Meadows, Ophuls argued that the Earth could only "carry" so much population. If economic growth and consumption continued to increase at current rates, he believed society would overshoot its growth limits. Collapse would then be inevitable.[24]

Maintaining biodiversity will prevent or at least postpone this fate, but biodiversity carries short-term utility as well. Such uses can be underscored in numerous respects, including the status of biodiversity as a viable component within the world economic system. As seen throughout the literature, for instance, biodiversity is a key tool within the pharmaceutical industry. Modern medicine has been greatly buttressed in recent years by applications derived from the diversity of organisms on earth, particularly the tropical plants that the Convention on Biological Diversity addresses.

At least 25 percent of the United States' prescription drugs today trace their active ingredients to plant matter, and bio-prospectors are continually discovering new applications. Utilization of the Ecuadorian rain forest plant *vehuco* for snakebites is one such example.[25] Rosy periwinkle, found in the rainforests of Madagascar, is another. It is used to treat childhood leukemia.[26] Many crops, furthermore, are genetically enhanced through wild species. While the mass-produced uniform crops of today are highly productive, their lack of genetic variation presents a handicap. These crops have decidedly more restrictive environmental requirements and are much more susceptible to pests and disease.[27] Genetically diverse wild species provide insurance against this threat.

Within this context of both short- and long-term utility, world leaders gathered in 1992 in Rio de Janeiro, Brazil, for the United Nations Conference on Environment and Development (UNCED), or as it is more popularly known, the Earth Summit. The largest gathering of state leaders ever, the Rio Summit sought to achieve multilateral agreements in five key areas, including the Rio Declaration, Agenda 21, Statement of Forest Principles, Climate Convention, and Biodiversity Convention. Despite high prioritization in a number of state delegations and vast press coverage worldwide, delegates came away with little more than non-binding recommendations. Development interests, particularly within the constraints of an overarching Northern-Southern hemisphere conflict, prevented any tangible results from emerging. As such, UNCED is widely recognized as failing to reach its initial promise. Instead, the primary legacy of Rio has been the extent to which the term "sustainable development" is now a contested terrain. Business interests, for one, co-opted its usage in the 1990s, often falsely equating qualitative development with quantitative growth.[28] If the international community is to overcome this ambiguity, it must first come to terms with the causes of biodiversity loss. This matter is undeniably complicated by disagreement over how important the species that are lost really are. It is this unknown, both in quantity and quality, that creates the political controversy.

The Causes of Biodiversity Loss

That said, the root causes of biodiversity loss demand attention. The three most damaging, in order of their negative impact, are habitat loss, invasive or non-indigenous species introduction, and general pollution. Each of these threats raise the question as to what role humanity should play in nature. Most significantly, continued encroachment on habitat the world over threatens the existence of countless species and subspecies. Nowhere is this more evident than in the tropics, the area where the vast majority of the world's species resides. Approximately three billion square miles of tropical forests remain to date. This region, which is equal in area to the forty-eight contiguous United States, is being deforested at the rate of 55,000 square miles each year, a tract approximately the size of Illinois.[29] Once an area is deforested, more than just trees are gone, though. The resources within and around the trees are gone as well. The above deforestation numbers also fail to incorporate the land that is degraded each year. Only deforested land, albeit at an acre per second, is included. A more accurate depiction of our overall predicament would consider soil erosion, silting of streams, climate change, desertification, salinization, and general loss of productivity. Put simply, much more than 55,000 square miles is actually lost each year. In fact, according to the World Resources Institute (WRI), tropical deforestation probably exceeds 130,000 square kilometers (80,600 square miles) a year.[30] Combining the deliberate ax, torch, bull dozer, and chain saw with unintentional erosion, salinization, compaction, nutrient depletion, and general pollution, it is easy to see that global land degradation has reached epidemic proportions.

Such losses highlight the fact that larger social and economic issues are intrinsically tied to biological diversity. This interdependence cannot be escaped, and any attempt at resolution must incorporate these perspectives. Still, over-development is not the root cause of biodiversity loss. It is the incentives behind excessive development (or perhaps better phrased, unnecessary growth) that policymakers must address instead. Effective management of biological resources requires recognition that blame for our current predicament exists at a number of levels. Most fundamentally, thoughtless, unsustainable consumption combines with basic human population growth to form the root causes of habitat destruction.

Both of these sources, though, come with significant political baggage. Consumption is best interpreted here as an issue of kind. Certain levels of consumption are of course natural, even welcome within an ecosystem. It is those that are thoughtless and emphasize quantity at the expense of quality that are destructive. Consumer demand of this kind as well as the more popularly cited population growth must be curbed, or carrying capacity limitations will set these ceilings for us. Population growth, similarly, must be couched within qualitative terms. Yes, quantitative growth is clearly a threat, but to gloss over basic inequity dimensions entrenched within this threat would be foolhardy. Too often, as Southern states are quick to point out, the North simply critiques Southern development strategies as unsustainable without incorporating these inequities. These are inequities, furthermore, whose sources should also be noted, as in many instances, Southern deficiencies stem from a combination not only of poor national management within the third world but also of the legacy of exploitation under Northern Hemisphere colonialism.

Like most global environmental problems, then, species loss is a function of both exponential developing-world growth *and* voracious industrial state consumption. This international dimension demands re-evaluation. At the same time, greater attention to the local needs of those who reside where the habitat destruction occurs, specifically the deforestation of tropical rain forests, is essential. Settlement policies, debt burdens, and land tenure agreements that support unsustainable agriculture within less-developed countries require re-examination. A two-pronged political dilemma then faces any attempt to resolve the biodiversity loss issue. Social concerns and economic desires fundamentally underlie biodiversity loss. In many cases, modern society fails to accurately assign the proper monetary value to natural resources, biodiversity included. As Jeffrey McNeely, chief scientist at the World Conservation Union, states with his colleagues, market systems erroneously emphasize productive values at the expense of consumptive and indirect values.[31]

True valuation must incorporate more than the productive uses of timber, ivory, medicinal plants, and the like within our environment. Consumptive uses such as firewood, wild game, and recreation are important as well, particularly on the local level. These values, though, do not pass through the market and often are ignored. Similarly, non-consumptive, indirect uses often fail to enter the economics valuation process. Indirect values of an ecosystem, granted, are the most difficult to estimate, yet watershed protection, photosynthesis, and climate regulation may be the most valuable of all variables in our worth equation. As such, biological resources are clearly undervalued in a monetary sense. Society, both local and global, must better match its social values to economic ones if it is to overcome this crisis. A growing literature recognizes this deficiency and identifies specific benefits and services that our environment provides, particularly in fostering life on Earth.[32]

A crucial part of such an adjustment requires targeting those who benefit from over-exploitation. This practice is indeed profitable for a minority group—in the short term. In the long term, of course, no one gains from biodiversity loss. Within a more myopic perspective, though, multinational logging companies who harvest their timber unsustainably effectively externalize their environmental costs. These costs fall to the local people of a region instead. Improving the fundamental understanding of biodiversity helps set the stage for reversing this inequity, but admittedly this knowledge only establishes a potential for reversing past exploitative practices. Monetary considerations re-enter the equation here, as substantial financial limitations continue to block attempts to link various state interests and local community interests together as one. Better funding mechanisms are needed. In addition, the focus of conservation must avoid the pitfall of becoming too narrow. Protective attempts must extend beyond single species, since entire ecological areas are more appropriate targets. In this regard, the wants and needs of specific local areas are a critical and often overlooked component of the biodiversity loss crisis. Such attention does not deny the international causes and implications that continue to shape this issue. It does re-orient the discussion, though, to a more appropriate focus. The relationships and dependencies existing within the state are as important as those between states.

POLITICAL SETTING

Sovereignty and Regimes

Three broad themes that emerge in the political literature concerning international environmental affairs will prove critical to this examination of biodiversity protection. Competing notions of sovereignty and regimes represent two distinct yet interrelated research programs that this section addresses together. Questions about domestic-international linkages are a second theme of importance, along with the scholarship on globalization that offers a unique twist here, albeit in a decidedly different theoretical context. Finally, the burgeoning literature regarding NGOs occupies the third main theme of political relevance regarding environmental protection, especially in the biodiversity arena. As with most issues of political science, and mirroring the ecological focus of this study, differing degrees of theoretical overlap exist here. One example is that scholars studying NGOs often employ previous scholarship on transnational regimes. Another is how the globalization literature acknowledges the role of regimes. Yet room for even further integration exists, particularly regarding substantive issues such as biodiversity protection.

Domestic politics have always been present in the international arena, but their influence is more pronounced with the global economy of today.

Nothing illustrates this better than the unique relationship between sovereignty and regimes. State sovereignty is defined as "the exclusive control over a given territory."[33] International regimes are the "implicit or explicit principles, norms, rules, and decision-making procedures around which actors' expectations converge in a given area of international relations."[34] The global environmental changes that now confront the world depend heavily upon the evolution of both these terms. This evolution, one must note, is more symbiotic than combative. Sovereignty and regimes need not square off in zero-sum conflict. Sovereignty is not fundamentally at odds with regime theory. While regimes clearly do alter sovereignty, the two can and, in fact, do co-exist. This is particularly true if one notes that sovereignty is "something variable that may be augmented or diminished."[35]

Indeed, the standard realistic conception of sovereignty as an unassailable principle within the international arena is, ironically, quite idealistic. As Karen Litfin, professor of political science at the University of Washington, asserts, sovereignty is a socially constructed institution that fluctuates depending upon place and time.[36] Similarly, as neo-realist Stephen Krasner effectively points out, sovereignty has never achieved a universally acceptable definition.[37] Sovereignty continually evolves, namely in relation to international security and order.[38] And perhaps the fluctuating nature of sovereignty, depending on both systemic international and specific domestic factors, allows for its future development according to conditions more hospitable to international cooperation. If one accepts this evolutionary status, moreover, future changes in the interpretations of sovereignty are allowed considerably more latitude. That is, changes need not be revolutionary. Adaptations could incorporate themes of pooled sovereignty that are nascent within the European Union (EU). A gradual movement in the direction of supra-nationalism is possible—instead of the standard either/or proposition regarding sovereignty or multi-state governance.

Before examining the degree to which recent scholarship reiterates this point, let us first examine the traditional realist approach to interpreting sovereignty. This school of thought argues that states have been the primary actors in the international system since the Peace of Westphalia in 1648.[39] Realist scholars of international relations from Hans Morgenthau to Kenneth Waltz and Robert Gilpin espouse the centrality of state sovereignty in international relations.[40] Despite challenges in the 1970s by liberal institutional scholars Robert Keohane and Joseph Nye and again by neoliberals of the 1990s such as Mark Zacher and Joseph Camilleri, who wrote extensively on multilateral institutions, realism remains entrenched as a viable theoretical approach due to its emphasis upon power as the guiding force in international relations.[41] This is critical in regard to sovereignty because, as stated earlier, sovereignty depends upon adequate power to exercise its authority.

Since the rise of interdependence theory in the 1970s, however, realist thought has come under increasing criticism. States alone cannot address the global pressures of today.[42] Nowhere is this deficiency more evident than within the transnational environment arena. One state is dependent upon another for acid-free rain, protective ozone in the upper atmosphere, and a biologically diverse biosphere. Liberal scholars recognize these dependencies. Some even challenge sovereignty outright. Zacher, for instance, contends that the six pillars of our "Westphalian temple" are withering under this onslaught.[43] Most importantly, Zacher notes the increasing levels of interdependence, rapid information flows, and growing instances of physical externalities that confront individual states in a manner that they are woefully unequipped to handle. Zacher even suggests that the Westphalian system is in such a state of disrepair that outright replacement of state sovereignty with internationally accepted norms and principles is inevitable.

The majority of the literature, though, still concludes that regimes will not supplant states as the primary actors on the international scene in the foreseeable future. Liberal institutional approaches such as Joseph Camilleri's do not sound the death knell for state relevance, but do challenge the "state's capacity to articulate and satisfy human needs" given the global pressures that continue to emerge.[44] Regimes and sovereignty will continue to evolve. Regimes will continue to instigate changes in sovereignty—but not usurp it. It is this emphasis on change that the transnational environmental literature seeks to exploit.[45] As international relations scholars Ronnie Lipschutz and Ken Conca assert, there is an emerging realization in international relations "that the world has entered an era of global social change running parallel to global environmental change."[46] Both sovereignty and regimes serve as essential components in this evolutionary process. In fact, despite obvious differences, regimes actually support sovereignty in key instances, and vice versa. Regimes both challenge state authority and assist in the preservation of sovereignty. Barry Hughes, professor of international relations at the University of Denver, for example, sees strong regimes as a support rather than a constraint for states simply because regimes encourage reciprocity.[47] While regimes do thrust another source of control upon an international stage where states were once the sole arbitrators, they also perform oversight and other functions that states increasingly find difficult to regulate on their own. As Karen Litfin, professor of political science at the University of Washington, states:

> Though these instruments (of international environmental regimes) constrain national prerogatives, and to that extent encroach on sovereignty, their benefits are seen as outweighing their drawbacks simply for the measure of protection they promise.[48]

While such approaches have done much to advance the study of global environmental problems, critics point out that regime theory is lacking in that it concentrates on the international or transnational level at the expense of domestic considerations. Paradoxically then, liberal scholars fall into the same trap as their realist brethren. They treat domestic factors as analytically separate from the international ones. Regime theory also fails to account adequately for the lingering effects of realism, specifically the continued preoccupation with short-term national power at the expense of long-term global security. States simply are not ready to sacrifice individual power, or potential power for that matter—even when long-term trends point to eventual power dissolution due to global environmental problems. Janet Abramovitz, senior researcher for Worldwatch Institute, typified this sentiment when speaking about the obstacles confronted at Earth Summit Plus 5 in the summer of 1997: "It's [the Climate Convention] a very national issue. Governments have not been very inclined to sign on to something that tells them what to do with their economic policies."[49] Domestic linkages to international environmental issues lie then at both the heart of the problem and the heart of any actions toward solution.

Domestic Links to International Relations

Yet domestic connections to the international are not made easily. As international relations scholars Elizabeth Economy and Miranda Schreurs note, these connections demand further empirical and theoretical research.[50] Even where the two levels are recognized in the same breath, most scholars, both realist and liberal, have been content to "debate the relative primacy of internal and external . . . rather than their connectedness."[51] Origins of this thought may be found in four general areas with a fifth research program of relevance, that which expressly emphasizes globalization theory, now emerging as well.[52] As early as the 1950s, functionalist theory, for one, asserted the role of spillover when transnational elites cooperated on a certain issue. Over time, provided the proper technical actor exists to assist in this transition, cooperation becomes habitual. Additional issues become negotiable. This is precisely the process that is unfolding today within the European Union (EU). French Foreign Minister Robert Schuman sought spillover when he first proposed French and German cooperation in the coal and steel industries in 1950. If historical enemies could find common ground on a resource vital to war, general political cooperation would be more likely to develop in the future. Peace would evolve. The last half-century has proven Schuman prophetic. The European Coal and Steel Community has withstood a variety of changes in that time—in name, membership, and ultimate objectives. While not a supranational entity, the EU does effectively incorporate pooled sovereignty, which allows for the daily interaction of the domestic and the international.

A second arena of note is how James Rosenau propelled this functionalist school of thought a step further in the 1960s with his call for an examination of the linkages between the domestic and international. Arguing that "non-members" of society interact with the official state members, that states are not the sole actors on the international stage, this second period of theoretical advance also served as a challenge to the dominant realist paradigm. Advocating a "conceptual jailbreak," Rosenau contended that national-international linkages should be treated as independent, not dependent, variables. Comparative specialists need not concentrate solely upon causes. International scholars need not focus entirely upon effects. Domestic policy researchers need not limit themselves to studying the motives of actors. Foreign policy academics need not dedicate themselves to only state capabilities. According to Rosenau, linkages among these subdisciplines should be examined as well, even though they are "extremely difficult to trace and measure."[53] Level of difficulty should not dissuade researchers from this agenda, for the merits to cataloging the contents "inside the black box" of the nation-state outweigh any conceptual (and counting) headaches that may arise. In some instances, moreover, linkages may actually simplify matters instead of complicating them.

A third avenue in which domestic-international relationships have been explored rests in the work of Robert Keohane and Joseph Nye, specifically their aforementioned *Power and Interdependence*. This seminal text and the complex interdependence it espoused serves as a foundation for contemporary linkage theory. With multiple channels connecting societies, a lack of general issue hierarchy, and decreasing emphasis upon military force, the potential impact of what previously were merely secondary influences is increased.[54] Complex interdependence raised the status of nonstate actors, revitalized international organizations, and introduced the concept of transnational coalition building for regimes. All this notwithstanding, one must still note that Keohane and Nye's work, as the authors themselves explain, does not systematically link process theory with domestic politics.

A fourth field of scholarly inquiry centers on international political economists who have adopted and adapted the domestic-international connections to various degrees, again relying heavily upon Keohane and Nye's complex interdependence. Throughout the 1980s, this theoretical approach permeated the foreign economic decision-making literature. Peter Gourevitch expounded on the degree to which new domestic policies are a reflection of international economic crises.[55] Ronald Rogowski, taking a slightly subtler angle, outlined the extent to which policy choice is spurred by international trade developments.[56] In both these cases, adequate understanding of domestic economic policymaking requires full comprehension of its linkages to the international level. Most research on domestic-international links into the twenty-first century continues to concentrate on this economic realm of complex interdependence.

Although some of the most interesting current research examines social and environmental dimensions, within this economic line of thought, a fifth school of interest emerges. This school, the globalization literature, continues paradoxically both to develop and expressly to challenge interdependence. The two terms "interdependence" and "globalization" often are used interchangeably in a popular context, but in theoretical terms they place much different emphasis on the state. Interdependence calls attention to the vulnerabilities that states face ecologically but also to the potential cooperation due to common interests. Globalization refers to the rapid communication and travel around the world today and its cultural implications, suggesting state sovereignty has eroded to an extent that new sources of order and authority are necessary. Dangers for unequal application of this order exist, most likely in a form repeating past transgressions against the developing world.[57] As the wealth of dependency literature demonstrates, such as Andre Gundar Frank's work on underdevelopment, industrial states reside at their current locus of power largely due to their exploitation of the third world.[58] Globalization could be merely the latest incarnation of the rich (an amorphous developed rich people instead of developed states) taking advantage of the poor.

Transcending this argument is the global-local tension that globalization scholars identify in the world today. Wolfgang Sachs, board chairman of Greenpeace Germany and senior fellow at the Wuppertal Institute for Climate, Energy, and Environment in Germany, for one, fears that global forces may homogenize our world to the detriment of cultures every-where. At the same time, Sachs also states that the global reach of many environmental issues today dictates the need for more than the traditional state-centric perspective. Looking to James Rosenau's consideration of global governance suggests that governance is possible without govern-ment, that more effective order can exist when states are not required to impose it.[59] The outstanding question then is whether this order is most effective when administered from below the state (locally) or above it (globally), or if some mix of the two along domestic-international linkages is possible.

The Role of NGOs

NGOs provide a promising medium to answering precisely this question due to their unique position both above and below the state. States are paradoxically both too big and too small when it comes to resolving transnational threats like biodiversity loss. As such, a new class of actors is needed.[60] NGOs represent this class. They are by no means the only set of actors that will determine the future of biodiversity protection. Yet, many now agree that they are the best positioned to ensure effective biodiversity protection. As University of Munich professor Ulrich Beck has noted:

> The future of ecological politics is not going to be decided in the circles of the established political institutions. It is precisely new coalitions among organizations, consumers, governmental agencies and NGO's [for which] we should be looking.[61]

Looking to the work of World Resources Institute's Gareth Porter and Janet Brown, we find three divisions of globally active environmental NGOs.[62] These divisions—international NGOs, large national NGOs, and think tanks—determine the appropriate parameters for future study of transnational biodiversity protection. With its headquarters in Amsterdam, Friends of the Earth International (FOEI), epitomizes the initial category. Some fifty-three national-based FOEI affiliates form a loose federation that meets annually to establish global priorities. Greenpeace represents the op-posite end of the international NGOs spectrum with its more centralized structure. Again, distinct national affiliates operate independently on the international level, but with decidedly more direction from their respective central headquarters. Large national NGOs such as the Sierra Club, Na-tional Audubon Society, and National Wildlife Federation within the United States represent the second category. Their primary focus is domestic—although various international interests within both their membership and their leadership have spawned an array of transnational programs addressing issues such as climate change and biodiversity protection. Think tanks such as Worldwatch Institute and World Resources Institute embody the third category. They publish reports on specific policy issues, often with the goal of directing the agenda during international conferences.

Clearly, then, environmental NGOs are a vast and diverse group. They are defined in broad terms by a 1950 UN resolution as "any international organization which is not established by intergovernmental agreement."[63] The *Yearbook of International Organizations* is even more general, stating that its editors accept any organization that declares itself to be an NGO as such. This analysis recognizes these broad categorizations but focuses upon not-for-profit organizations, even though businesses are included in the United Nations' definition of general NGOs.[64] Even with this limitation, an enormous number of groups are found. Indeed, NGOs command attention due to their numbers alone. While the rate of increase in the number of states appears to have leveled off and the number of intergovernmental organizations (IGOs) is actually in decline, the last twenty years have seen a veritable explosion in the number of NGOs on the international stage. Environmental NGOs, it should be noted, represent a significant portion of this development.[65] American University's Paul Wapner, for one, believes that over one hundred thousand different environmental NGOs exist.[66] The great majority of these environmental NGOs, as identified by the World Resources Institute, began in the 1980s. Not coincidentally, the emergence of NGOs on the international scene coincided with a growing emphasis on the global environment in international affairs. The twenty-first century, with rampant globalization and a possible return to a multi-polar power structure, moreover, presents a climate conducive to even more intensive examination of this phenomenon.

Nongovernmental organizations pursue their objectives in a number of legal and policy arenas. Specific activities here range from pollution control to wildlife preservation to poverty minimization. Methods vary from research to education to lobbying to civil disobedience and, at times, actual implementation. Ideological orientations cover a spectrum from realist greens to feminist environmentalists to deep or spiritual ecologists. Some NGOs such as the Sierra Club, Friends of the Earth, or Greenpeace emphasize the grassroots. Others like World Wildlife Federation and Natural Resources Defense Council are more traditional, employing the assistance of big business in their strategic objectives. With all this variety, not surprisingly, differences exist within the literature with regard to the most useful theoretical approach for studying NGOs. There are those that concentrate upon NGO expertise as an asset to state decision making, those that emphasize NGOs as an alternative power source in transnational environmental issues (either as a linkage agent between domestic and international concerns or as an outright challenge to state control), and those that focus upon construction and maintenance of a global civil society.

First and foremost within the literature, scholars such as Porter and Brown laud the role of expertise. The degree to which NGOs ply expert knowledge for complex scientific predicaments makes them critical international players. NGOs, quite simply, translate knowledge into action. Ambassador Richard Benedick, who led the United States delegation in negotiating the Montreal Protocol, points out the contributions of the Natural Resources Defense Council (NRDC) in finalizing that treaty. NRDC, with its compilation of state reduction proposals for ozone depleting chemicals, set the 1990 negotiating table—one that led to significant reductions in the Montreal Protocol targets. Peter Haas, professor of political science at University of Massachusetts Amherst, also addresses expertise by stressing the phenomenon of epistemic communities conducive to policy resolution, an emerging trend within the discipline.[67] Barbara Bramble and Gareth Porter, utilizing their own NGO backgrounds, similarly acknowledge the role of research credibility as a major NGO contribution.[68] Also note worthy is praise by academic Lamont Hempel, that international environmental NGOs are usually better prepared than governments to implement studies of environmental protection.[69] Added together, these examples of expertise are all the more effective when coupled with the potential for innovation, a quality many more static state organizations cannot achieve. The contributions of this combination cannot be overstated. As Alison Jolly states:

> The enormous importance of NGOs lies in their expertise. It is they who know scientists, both foreign and local. It is their people who have slept in the woods, bird-watched at dawn, negotiated at noon with the provincial politician, and exchanged oratorical speeches of friendship with the local chief under the moon, washed down in local beer.[70]

Porter and Brown also introduce an important corollary here. Some of the best examples of NGO success are found in their degree of influence over a major actor. That is, expertise can be particularly effective when NGOs target key states like the United States. The success of the U.S. Clean Air Coalition illustrates one such case, as they effectively lobbied for regulation of chlorofluorocarbons (CFCs) in the late 1970s and early 1980s. Another example is the activity by the World Resources Institute

(WRI), World Wide Fund for Nature (WWF), and the Environmental and Energy Study Institute within the United States.[71] These NGOs worked closely with the pro-CBD biotechnology firm of Merck and Genentech in lobbying the first Clinton administration. In 1993, with their draft of an interpretative statement, they helped convince President Clinton to reverse the previous administration's position and sign the treaty—although the United States remained a non-ratified observer throughout the remainder of the two Clinton terms and into President George W. Bush's administration.

Leadership within the United Nations as well as activity around it supports this assessment, that NGOs offer expertise many states actively apply. Mostafa Tolba, executive director of United Nations Environment Programme from 1976 to 1992, identifies NGOs as one of key actors that found influence during negotiations for treaties addressing ozone depletion, global warming, hazardous waste trade, and loss of biological diversity.[72] Ranee Panjabi of Memorial University in Canada, similarly, stresses the role of NGO expertise in her comprehensive examination of the United Nations Conference on Environment and Development (UNCED), detailing both major accomplishments and failures as they relate to both the Convention on Climate Change and the Convention on Biological Diversity.[73]

The second avenue of scholarship on NGOs addresses the degree to which they challenge state authority, or at least provide alternative power sources. NGOs at times wield transnational power because they are not limited by parochial allegiances.[74] They provide a medium for international interests as well as domestic interests. This is not to deny the presence of South-North rifts, particularly at key international conferences such as the 1992 United Nations Conference on Environment and Development. As Porter and Brown demonstrate, northern NGOs are often better prepared, better financed, and better represented.[75] Still, despite developing and developed state emphasis differentials, objectives often do transcend narrow national interests—and at times the North-South division as well. Indigenous opposition to rainforest destruction in Brazil linked with United States–based NGOs, for example, to halt various construction projects within the Amazon in the 1980s. This coalition also convinced Brazilian president José Sarney to scrap tax incentives for agriculture that encouraged deforesting the area. Similarly, it persuaded the United States Congress to help fund reserves allocated toward conservation efforts.[76]

Political economist Thomas Princen and political scientist Mathias Finger, including their two additional contributors, provide a strong foundation for future research in this regard with their examination of four environmental case studies: the Great Lakes water negotiations, the ivory trade ban, Antarctic environmental protection, and the United Nations Conference on Environment and Development. Princen and Finger do not argue overtly that NGOs challenge state authority, but they do outline a power source acting independent of the state. The authors clearly follow in the tradition of Rosenau's call for linkages between domestic and international. Adapting the theoretical lenses of political bargaining and social movements, they find distinct applications for environmental NGOs:

> NGOs fill a growing diplomatic niche . . . built not on territory and natural resources nor on the ability to gather taxes and marshal armies. Rather, it is the influence achieved by building expertise in areas diplomats tend to ignore and by revealing information economic interests tend to withhold.[77]

Both these functions of expertise and alternative power sources draw heavily upon the aforementioned neo-realist versus neo-liberal debate that permeates all of transnational relations. In studying NGOs in the 1960s, most scholars simply chose one theoretical perspective or the other. Realists argued that NGO growth reflected hegemonic stability.[78] It was a function of inter-state behavior. Liberals countered that NGOs continued to clip state authority in transnational affairs, that "sovereignty was at bay."[79] Still, even the liberals, cloaked in neo-liberal garments, measure effectiveness of NGOs according to their degree of influence within states.

While much contemporary research continues in this vein, the most interesting work does not follow this either/or format. Instead, as seen in the third function of NGOs described within the literature, NGOs both construct and maintain our evolving global civil society.[80] This phenomenon continues to develop today, as evidenced by the groundwork laid at UNCED in 1992. As Lorraine Elliott of Australian National University in Canberra contends, the series of voluntary agreements achieved by states is notable but the real shift since UNCED is the movement toward a global civil society, thanks in large part to the NGO contingent at Rio. Over thirty thousand individuals representing a litany of NGOs participated in conferences held in conjunction with the official state meetings of UNCED and produced thirty alternative treaties. Unlike their official government counterparts, real progress was made.[81] They set a precedent in 1992 to do more than merely provide technical assistance, showing that NGOs act as conduits for ideas. They provide global public goods.[82] They create the institutional capacity for environmental protection. As NGOs persuade, bargain, and coerce, they contribute to development of a global civil society. They demonstrate the degree to which the environmental problems of today have become "part of the common experience of humanity."[83] Working with local, grassroots groups, NGOs then mobilize collective action.

In his analysis of the World Bank and the environmental degradation it has sparked in the developing world, Bruce Rich, program manager of the international program at Environmental Defense, emphasizes precisely this characteristic of NGOs. Rich believes that NGOs "articulated local social and environmental interests on a worldwide scale," beginning in the 1980s.[84] He explains how NGOs combine the contributions of knowledge and alternative power sources to this third arena of global civil society to force actual governmental policy changes. His description of changes in the Brazilian constitution, particularly provisions that recognized Indian rights to land and general Brazilian rights nationally to a healthy environment, is a good example. Rich also credits NGOs for the creation of various environmental agencies within the national government of Brazil and the establishment of specific environmental impact procedures that encouraged broader public participation. All this was possible, Rich asserts, thanks to a nascent global civil society that is shaped in large part by NGOs. As Rich explains:

> Armed with lap-top computers, faxes, and modems, NGOs in the Amazon have formed powerful, responsive networks with national groups in São Paulo and Rio, and international NGOs in Washington, Berlin, and Tokyo. They influenced — and are continuing to influence — not only Brazilian government policy, but also the flow of public international finance, through a kind of global electronic polis.[85]

Rich identifies the most powerful contribution of NGOs as their ability to link local impacts with global actions. This is his focus in examining the role of NGOs in negotiating a more equitable and ecologically benign international lending process within the World Bank. Citing numerous examples of discursive NGO lobbying, he describes a World Bank that, while only making procedural changes to date, has taken steps to re-define sustainable development in the context of its lending practices. Despite these advances, further opening of the notoriously secretive proceedings in the World Bank is needed before truly sustainable strategies are implemented.[86] The Bank must come to a more complex understanding of those who live in an area of "development" than the current dualistic consideration of either "beneficiaries" or "PAPs" (project-affected populations). If not, future disasters on the order of the transmigration which

Rich critiques in Indonesia and the massive hydroelectric developments in Thailand and India are likely.[87]

Like Rich, Paul Wapner, professor in the international service program at American University, focuses upon the civil society dimension of NGO activity in his analysis of environmental activism.[88] Wapner argues that, while conventional, state-centered interpretations are not wrong per se, they are incomplete. Environmental NGOs, as seen through Wapner's analysis, provide an ideal channel to link the domestic and international. Indeed, the prospects for biodiversity protection rest firmly with effective linkages between state and international concerns.

What Comes Next

Despite the work outlined above, the NGO field of study remains in its early stages of development. As Harvard University's Sheila Jasanoff contends, "systematic assessment" of the role that NGOs play in environmental decision making remains conspicuously absent in the studies to date.[89] The intersection of the regimes and sovereignty literature with the domestic-international linkage literature combines with the above scholarship on NGOs to present a unique opportunity for scholars examining transnational biodiversity protection. Research does appear to point toward the potential for NGOs in furnishing much-needed scientific expertise to the diplomatic table. Still, contributions are needed regarding the role that NGOs play in environmental decision making—and specifically how NGOs go about performing that role. This gap demands attention, particularly since NGOs relate to governing institutions and serve as crucial domestic linkages to the international.

As outlined earlier in this chapter, the regime literature to date demonstrates that the veil of sovereignty is thinning—but fails to come to terms with the continued obstacle of realist state power. Part of this omission may be explained by neglect of domestic influences, a deficiency that the second research program from above, domestic-international linkages, attempts to correct. This research program, however, also remains somewhat ambiguous in terms of clearly establishing these links. It is here that NGO research attempts to fill the gaps by providing a tangible medium for the links between the domestic and international. To date, this NGO research explains much about these interactions, but little about how they develop—and more expressly how they can be improved. More research is needed in this direction, particularly with regard to strategic effectiveness. What makes NGOs effective, or conversely ineffective, is a crucial political question for future biodiversity protection efforts.

Table 2. International Biodiversity-Related Conventions

	Focus	Number of state parties	Date negotiated	Entered into force	U.S. status
Convention on Biological Diversity (CBD)	Global biodiversity	187	1992	1993	Signed in 1993 but not yet ratified
Convention on International Trade in Endangered Species (CITES)	Trade in international wildlife	162	1973	1975	Ratified in 1974
Convention on Migratory Species (CMS or Bonn)	Migratory species	81	1979	1983	Neither a party nor a signatory
Convention on Wetlands (Ramsar)	Originally only waterfowl, but today wetlands more generally	136	1971	1975	Contracting party since 1987
World Heritage Convention (WH)	Links cultural and natural heritage	176	1975	1975	Ratified in 1973

Note: Based on data through June 2003.

INTERNATIONAL BIODIVERSITY AGREEMENTS

Introduction

Several biodiversity protection efforts exist at the international level. Five of particular note are the Convention on Biological Diversity (CBD); the Convention on International Trade in Endangered Species of Wild Fauna and Flora (CITES); the Convention on Wetlands of International Importance, Especially as Waterfowl Habitat (Ramsar); the Convention on Preservation of Migratory Species of Wild Animals (CMS); and the Convention Concerning the Protection of the World Cultural and Natural Heritage (WH).[90] Table 2 summarizes each of these.

In an effort to develop synergies among these conventions, the Nineteenth Special Session of the UN General Assembly in June 1997 recommended collaboration of the various information capacities here toward further implementation of the Agenda 21 and its goal of sustainable development. The hope was simply that the various linkages inherent within these conventions would be exploited. As such, these five biodiversity-related conventions exchange information on behalf of their party members, facilitated through a joint Internet presence hosted by the CBD.[91] Out of these five conventions, the two most often targeted agreements by NGOs are CITES and the CBD.

The Convention on International Trade in Endangered Species

The 1973 Convention on International Trade in Endangered Species (CITES) comes the closest to CBD in terms of international impact, but it is by definition limited to known species that are officially recognized as endangered—and only within the context of trade. With its administrative body in Geneva, Switzerland, CITES has 162 state parties within its membership. This includes the United States, which was actually one of the ten original party states when the treaty first went into force in 1975.[92] Since that time, twelve Conference of Parties have been held, with the most recent in November 2002, in Santiago, Chile, attended by some 1,200 participants from 141 governments as well as numerous observer organizations.[93] These meetings are used to amend and update the three appendices of species categories that form the convention. Appendix I is the most protective, prohibiting all international commercial trade in threatened species. Appendix II also provides a degree of protection, as it regulates trade in species not currently threatened with extinction but susceptible to such a threat if trade were to be unregulated. Appendix III is simply a list of species that countries stipulate as already protected within their own borders.

CITES is also noteworthy in the extent that it uses relatively strong language limiting NGO participation, language that has become the model for subsequent treaties.[94] NGOs retain a second-class status—and may even have their observer position rescinded in a given meeting if one-third of state parties concur. Another example of NGO second-class status is the fact that they are banned from both participation and observation of any CITES budget committee sessions. Despite this secondary status, though, NGOs are able to exploit the informal channels that surround an agreement once it enters into force. John Lanchbery, of the London-based Verification Technology Information Centre, for one, finds that NGOs such as WWF-International and the NGO-IGO hybrid IUCN have played critical roles in the development of CITES. Outlining thirty-four international agreements on fauna and flora (he includes regional as well as global agreements) beginning as early as the late nineteenth century, Lanchbery concentrates on CITES, Ramsar, and the CMS.[95] Lanchbery believes implemen-

tation review to be the real area of potential influence for NGOs. Although
he also recognizes the fundamental role of raising awareness of the issues,
Lanchbery emphasizes NGOs for their ability to exploit the informal channels that surround the adjustments to an international agreement over
time. They shape how an agreement evolves.

The Convention on Biological Diversity

The Convention on Biological Diversity (CBD) represents the first and
only international treaty protecting total species diversity. It also attempts
to incorporate long-standing NGO themes such as agricultural problems,
external debt, distribution of wealth, and general development critiques.[96]
Signed initially by 157 states, some 187 states are now full-fledged members given their ratified status.[97] The Convention entered into force in
December 1993 after being ratified in record time as the first global treaty
to "explicitly take a comprehensive, ecosystems-based approach to the
protection of biodiversity."[98] It is a product of the 1992 UN Conference
on Environment and Development (UNCED) in Rio de Jainero, Brazil,
where representatives (including 108 world leaders) gathered from 178
different countries. The Earth Summit, as it is popularly known, represented the largest gathering of world leaders in history. All told, some ten
thousand diplomats were present. Four main agreements in addition to the
Biodiversity Treaty were produced: Agenda 21, the Rio Declaration, a
Statement of Forest Principles, and the Climate Change Treaty.[99]

Despite six Conferences of Parties in almost ten years, though, the potential of the CBD continues to be stymied by United States' intransigence.
Action (perhaps more accurately stated as lack thereof) was limited to essentially a series of stocktaking meetings in June 1997 at Earth Summit
Plus 5.[100] This special session of the United Nations General Assembly,
with seventy heads of state and representatives of 199 countries and intergovernmental organizations, concentrated upon climate change but set no
targets or timetables. Moreover, the CBD was not specifically addressed
that summer, possibly underscoring the degree that obstructive domestic
considerations are even more prevalent in this issue area than that of climate change.

With regard to the Climate Change Convention, President George W.
Bush publicly opposes the Kyoto Protocol, believing that the 7 percent re-
ductions from 1990 emissions levels would "cause serious harm to the
U.S. economy."[101] Admittedly, the reduction in emissions would be costly,
since the United States is polluting much more today than it was in 1990.
Indeed, at the current rate of increase in emissions, the United States

Figure 1.2. Neglecting the precautionary principle. SOURCE: Kirk Anderson <www.kirktoons.com>.

would actually require a 34 percent reduction by 2010. This is but one more example of how politicians can twist statistics. We would be wise to heed Mark Twain's age-old warning in this regard, that there are three types of lies: lies, damn lies, and statistics. A suspicious mind, for example, might ask why the Germans were so keen on using 1990 as the baseline? It does, after all, allow them to include East German emission rates, thus making it easier to reach targeted reductions. None of these caveats should be construed as condoning the United States snub of the rest of the world, even for financial reasons. In fact, economic reasons should actually encourage United States participation. Simply requesting further "study" of this "discernible human influence" on climate change throws caution to the wind and risks a financial soaking in both a figurative and literal sense.

In any case, international progress has stalled accordingly. President George W. Bush's predecessor, President Clinton, signed the treaty, but never seriously sought ratification, because he received substantial pressure against it from both industrial leaders and Congress. Another notable stumbling block is the fact that developing countries such as China and India are exempt from the stringent targets. Some Kyoto opponents fear that production would likely shift to these states, creating possibly even higher levels of overall pollution due to the lower emission standards in those countries. The U.S. Senate is on record against the Kyoto agreements and has voted 95 to 0 against approval of any climate agreement that allows exemptions for developing states—a feature that the December 1997 treaty contains. Prospects remain grim, then, for the basic fact that Kyoto is set up as a dual-trigger treaty. As stated in article 25 of the Protocol, to enter into force, Kyoto requires 55 parties to the Convention to ratify the Protocol, including states representing 55 percent of the carbon dioxide emissions based on 1990 levels from the Annex I parties. As of June 2004, the protocol had more than met the first requirement with 122 accessions, but remained stuck on the second with only 44.2 percent of emissions accounted for by those Annex I parties. Either Russia at 17.4 percent or the United States at 36.1 percent is needed for the treaty to enter into force.

Unlike the Climate Change Convention, the CBD is in force despite lack of U.S. support. But similar developed state versus developing state divisions do exist and constrain its practical effectiveness. The essentials of the CBD may be broken into four critical components. First, biotechnology and pharmaceutical interests must share their profits with local entities. Article 19 of the Convention clearly stipulates that contracting parties of the Convention must come to prior agreement on the "handling of biotechnology and distribution of its benefits."[102] This is a critical point, since developing countries hold approximately four-fifths of the world's biological diversity. This, of course, also raises a red flag for groups within the United States, as some feel such language places property rights under siege. The term "benefit sharing," in particular, appears to challenge United States arguments surrounding intellectual property rights. Second, signatories must develop national plans for protecting both individual species and the overall habitat itself. This includes both the in-situ conservation addressed in Article 8 and the ex-situ conservation discussed in Article 9. Third, wealthy states must assist in funding these national plans. As Articles 20 and 21 outline, "the developed country Parties shall provide new and additional financial resources to enable developing country Parties to meet the agreed full incremental costs to them of implementing measures."[103] Fourth and finally, such financial support from the developed states guarantees commercial access.

Significant social and economic hurdles confront each of these four steps. Former President Clinton recognized the economic aspects of this in his attempt to attach side agreements to ensure United States ratification early in his first administration. These unilateral interpretations would protect patents and commercially enterprising research, areas that included industries directly affected by the CBD. Still, these overtures were not enough to push through ratification in the Senate. And President George W. Bush's administration has taken no action to date. Economic concerns remained, specifically the codification of the developing states' sovereign right to their genetic resources. Such language contrasted starkly with United States desires to protect existing intellectual property rights. Previous international agreements like the non-binding 1983 UN Food and Agriculture Organization's Undertaking on Plant Genetic Resources recognized intellectual property rights and thus supported free access to genetic material. The CBD dramatically reversed this practice—although some attempt is made to reconcile developing states' sovereignty and developed states' intellectual property rights in Article 16.

Thus, with regard to the CBD, international cooperation per se is not the fundamental problem. A virtual consensus exists on the international level. The key remaining obstacle is a national one, specifically domestic politics within the United States.[104] In June 1993, President Clinton reversed President George Bush's stance on the treaty by signing the document. In the summer of 1994, on the heels of Clinton's signature the previous year, the Convention on Biological Diversity appeared to be headed for certain ratification in the United States Senate. The Foreign Relations Committee demonstrated bi-partisan support in passing their approval to the floor with a 16 to 3 vote. That changed when Republican presidential hopeful, Senator Bob Dole, sent a letter to the committee chair expressing reservations about the treaty. Dole's letter, co-signed by thirty-four GOP senators, called for postponement of debate at least until after the 1996 elections. Much of this rationale was clearly motivated by presidential campaign posturing. Yet, one must not ignore the fact that ten of the eleven environmental bills submitted to Congress in 1994 failed, including a February 1994 proposal to elevate the EPA to cabinet rank.

With the end of the Clinton Administration and President George W. Bush taking the helm, the United States remains simply a non-ratified observer, with no sign of movement on the issue. More than mere partisan politics is needed to explain this occurrence.[105] A deeper undercurrent of pocketbook-minded public opinion may be identified, one that also explains reactions to previous environmental regulation attempts within the United States. This same tide of economic concerns helps explain President George Bush's stance at Rio in 1992. For the first half of his administration, President Bush was counted in most circles as a member of the environmental camp, supporting the Clean Air Act Amendments of 1990 and a host of other legislation. As he geared up for the 1992 election, however, Bush reversed his environmental stance. Going against the wishes of his EPA administrator, William Reilly, Bush heeded the advice instead of Office of Management and Budget (OMB) director Richard Darman and adopted a defiant, anti-regulator attitude at the Earth Summit. This positioned the United States as the only major industrial country *not* to ratify the treaty. As such, it wields only observer state authority and is unable to assume the proper leadership role its hegemonic status would otherwise suggest. The CBD, without support from the remaining world superpower, is effectively emasculated.

EFFECTIVE TRANSNATIONAL BIODIVERSITY PROTECTION

NGO Effectiveness

Many cases of NGO influence within the transnational environmental policy arena exist. Examples include Rainforest Action Network (RAN) with their initiative against Amazon deforestation activities precipitated by cattle ranching for Burger King and its hamburgers, Earth Island Institute's (EII) lobby for dolphin-safe tuna, Environmental Defense's campaign against McDonald's use of polyurethane wrapping, and Friends of the Earth's (FOE) stance in support of CFC alternatives. Despite these examples of influence, though, effectiveness itself remains a remarkably difficult concept to measure. Too often, the causal links from an action to its effects are frustratingly ambiguous. Identifying the precise NGO action that changes behavior in the various targeted entities (including both states and civil society) is complicated by both the nature of the issue and the number of actors involved in it. Transnational biodiversity is complex and interdependent—and attracts many interests. NGOs are not the only actors involved in transnational biodiversity protection. Thus, when progress is made on this issue, the assignment of effectiveness because of one actor or another can be difficult. Despite their best intentions, a merely spurious correlation may exist between NGO efforts and any effective protection initiatives that develop.

Within this context, considerable uncertainty surrounds measurement of effectiveness. Oran Young, a noted regime theorist now at the University of California at Santa Barbara, and Marc Levy, associate director for science applications at Columbia University's Center for International Earth Science Information Network (CIESIN), see effectiveness as an elusive concept and open to a number of different interpretations.[106] Young even laments in his examination of regimes that "there is no simple and straightforward way to define effectiveness."[107] Indeed, with divergent normative, scientific, and historical judgments, defining effectiveness is a difficult task. Legal interpretations of effectiveness represent one option to solving these obstacles by sidestepping pressures for normative assessment. Noting the degree to which contractual obligations are met allows measurement of effectiveness with decided operational clarity. Simply complying with rules would constitute effectiveness.

Of course, this presents practical problems for this study as the major international agreement concerning transnational biodiversity, the CBD, is an inoperable medium for protecting biodiversity by virtue of the United States nonparticipation in the treaty. Without the United States as a party to this convention, the CBD will never be able to protect transnational biodiversity adequately. NGOs recognize this deficiency and since 1992 have attempted to convince the United States to become a fully ratified member. As Carroll Muffett, International Programs director at Defenders of Wildlife asserts, "Clearly the CBD hasn't achieved what people thought it could achieve. [But] you have to view it as a framework treaty. There is some meat in the agreement itself, but without further protocols the CBD doesn't have a lot of bite."[108] And there are some notable successes here such as the bio-safety protocols completed in January 2000. Still, prospects for fulfilling the original intent of the CBD remain fuzzy, and NGOs have broadened their target areas accordingly.

It is in this same spirit that examinations of effectiveness must be enlarged as well. NGOs, like the regimes addressed in Oran Young's edited work, can be effective without noting rule compliance along treaty lines. As a broader measure, political effectiveness considers precisely this and incorporates the most intuitive, appealing sense of effectiveness. By taking a problem-solving approach, political effectiveness "centers on the degree to which an [NGO] . . . eliminates or alleviates the problem."[109] Measures of political effectiveness target changes in the behavior of actors and interests, and recognize action toward achieving objectives. The objective itself does not have to be met. This definition, as Young and Levy note, does not exclude considerations of whether problems are solved or a high enough compliance rate is met. It only changes the focus to behavior. As Young and Levy write, "activities that move the system in the right direction, even if they fall short of full compliance, are signs of effectiveness."[110] David Victor, Kal Raustiala, and Eugene Skolnikoff, showing a mixture of applied, legal, and political stripes due to their respective positions, also focus upon behavior changes as the best measure of effectiveness. Opting not to emphasize compliance, these researchers contend that high rates of compliance in international environmental law actually say little substantively about effectiveness. They argue that compliance is simply a means to achieve effectiveness. It is not an end in itself. While studies of compliance are beneficial, they often leave out the fundamental component of behavior change and fail to recognize when standards within a treaty are "too weak, too strong, inefficient, or completely ill conceived."[111]

In their study of institutions, scholars Peter Haas, Robert Keohane, and Marc Levy also identify a political dimension for effectiveness. Haas, Keohane, and Levy identify three prerequisites for effective action on environmental issues to occur, namely "high levels of governmental concern, a hospitable contractual environment, and sufficient political and administrative capacity in national governments."[112] Calling these conditions the "three *c*'s" (concern, contractual environmental, and capacity), this analysis may easily be applied to NGOs. Indeed, NGOs, while offering no magical formula, do participate at times in all three of these activities. NGOs help set agendas. They help shape international policies. And they help construct national responses to these international policies. NGOs do this in a number of ways. As a primary entity in the construction and maintenance of regimes, NGOs often adapt and adopt the six contributions that Young and Levy identify for international environmental regimes. NGOs act as utility modifiers by identifying the costs and benefits of various actions. NGOs improve cooperation. NGOs assign authority. NGOs foster learning. NGOs define and redefine roles, explaining the position that various actors take on an issue.[113] And NGOs serve as the impetus to internal realignments on an issue, whether this is by creating new constituencies or simply changing the relative strength of various factions regarding an issue.

In this context, this book identifies four key indicators to determine NGO effectiveness in meeting the three fundamental linkages of domestic-international, ecological-economic, and short-term–long-term considerations. As Donald Wells of West Georgia College explains, environmental NGOs are most effective when they define issues, mobilize individuals or groups of individuals to address a problem, and articulate that issue interest to government agencies.[114] These three conditions frame the context for potential effectiveness in this book. They essentially outline a *symbolic* effectiveness that is of direct utility for this study. Along with these conditions, a still higher gradation of effectiveness, one that measures the aforementioned *political* effectiveness, is measurable. Such an indicator recognizes more than mere activity by an NGO. It recognizes that more than presentation of papers and writing of books is at work. It recognizes that NGOs contribute to the second and third *c*'s of Haas, Keohane, and Levy, namely that NGOs provide input on international policies and that NGOs provide input on specific state reactions to those policies. This political measurement notes when NGOs are taken seriously by states or inter-national organization authorities and when they are admitted into the inner sanctums of decision making.

This analysis, then, incorporates both symbolic and political effective-ness in analyzing the ability of NGOs to establish the three requisite link-ages. An NGO is defined as effective at establishing a linkage when it meets each of the following four indicators of effectiveness:

1. The NGO defines an issue . . . and its definition prevails.
2. The NGO mobilizes staff or membership around this issue.
3. The NGO articulates that issue to key policymaking access points.
4. The NGO assists in implementation.

Again, it cannot be overstated that this fourth point is critical to effective-ness. As the literature suggests, implementation is the basic process that turns commitments into action.[115] That is, implementation influences, shapes, or even changes policy or the policymaking process in some form, by entirely re-defining a policy or reconfiguring the parameters in which an issue is considered. Such an approach is also useful here as it examines changes in behavior as Young does in his examination of regimes. While ef-fectiveness as defined above is not as stringent a measure as success of spe-cific environmental agreements, it does recognize the role that NGOs play as utility modifiers, enhancers of cooperation, bestowers of authority, learn-ing facilitators, role definers, and agents of internal realignments.[116]

NGO Linkage Tasks

The central research question of this book asks what enhances the ability of an NGO to protect transnational biodiversity. Three fundamental linkages are notable here. NGOs that are able to make these linkages will be more effective than those that do not. It should also be noted that, much like the issue area of this study, none of these linkages operate in a vacuum. Each displays highly advantageous synergies with the other linkages. Domestic-international, ecological-economic, and short-term–long-term considera-tions are all intrinsically linked together, and the most effective NGO strate-gies exploit these relationships. These linkages are expressly stated as the following:

Linkage 1.
To achieve effective transnational biodiversity protection, environmental NGOs must link domestic concerns with international concerns.

For example, NGOs must equate Northern concerns about species protec-tion and intellectual property rights with Southern development desires and insistence on retaining resource availability, if they are to effectively protect transnational biodiversity.

Linkage 2:
To achieve effective transnational biodiversity protection, environmental NGOs must link ecological concerns to economic and social concerns.

That is, effective transnational biodiversity protection is only possible when NGOs address the economic and ecological needs of a given community as fundamentally interrelated issues. They must recognize the complex and interdependent nature of the issue. Of course, this linkage overlaps with the previous one in that these economic-ecological connections must also be made within the context of communities beyond the immediate physical location of a resource.

Linkage 3:
To achieve effective transnational biodiversity protection, environmental NGOs must link short-term concerns with long-term concerns.

Addressing only short-term interests, or inversely, only long-term interests, is not enough. Effective transnational biodiversity protection demands both short- and long-term perspectives. Again, significant overlap with each of the above linkages occurs here. Short- and long-term considerations incorporate ecological and economic as well as domestic and international linkages.

A FRAMEWORK FOR NGO EFFECTIVENESS

NGO Strategies

Asking precisely how NGOs help build the next ark, this book looks at both mainstream and participatory strategies. Within the three general approaches to NGOs outlined earlier in this chapter, the research indicates fundamentally divergent approaches in strategic emphasis that range from political networking, which employs the work of noted sociologist James Coleman's conception of social capital, to highly charged confrontational tactics such as those often applied by NGOs such as Greenpeace, Earth Liberation Front (ELF), or Sea Shepherd Conservation Society.[117] Dave Foreman's account of Earth First! illustrates the degree to which radical organizations can persuade more traditional groups like the Audubon Society and the Wilderness Society to readjust their priorities. The movement to preserve old growth in national forests is a case in point. Opposition to logging began in fringe groups such as Earth First! before jelling into a na-

Table 3. Typology of Mainstream and Participatory Strategies

Strategies	Action focused	Data focused
Mainstream	Lobbying	Scientific research
	Litigation	Acquiring/managing property
		Monitoring agreements
Participatory	Grassroots networking	Community education

tionwide awareness. The radical opposition shifted the debate from what was previously a concern with scenery and recreation to the more formidable structure around the biological diversity needs prevalent at the end of the twentieth century. Quite simply, EarthFirst! jump-started a reframing of the question of wilderness preservation. What was once purely aesthetic became ecologically utilitarian.[118]

Building on these differences in strategic approaches, this analysis identifies seven key strategies that NGOs utilize most regularly. These are categorized as mainstream strategies and participatory strategies in table 3, each of which has the typology of action-focused and data-focused items. Admittedly, a high degree of overlap exists among the action-focused and data-focused categories in table 3. These are not discrete boxes and the lines would be best drawn as dotted, if at all. Moreover, the underlining argument in this book is that mainstream and participatory strategies overlap as well. Nevertheless, it can be useful to conceptualize strategies in these divisions. The first strategy, mainstream lobbying, entails lobbying of legislative and executive agencies both at home and abroad as well as at international negotiations and conferences. It is perhaps the most common NGO strategy, at least in the public mindset.

For example, at times NGOs draft convention texts prior to conference meetings in an effort to set the agenda. Only the International Union for the Conservation of Nature (IUCN) has been successful with this strategy to date in providing an NGO-related proposal that served as the actual starting block for negotiations—and even here there is a key qualification, as IUCN is a hybrid NGO-IGO. Still, their draft served as the basis for the 1972 Convention Concerning the Protection of the World Cultural and Natural Heritage in Paris. The 1973 Convention on International Trade in Endangered Species (CITES) is also based on an IUCN draft. With 76 states, 104 government agencies, 720 NGOs, 35 affiliates, and some 10,000 scientists, the IUCN is the largest conservation-related organization in the world. Its general assembly of delegates meets every three years and has had significant impact upon wildlife conservation and species protection.

One note of further explanation on my word choice is necessary here. Environmental NGOs often use the term "advocacy" instead of "lobbying," because "lobbying" carries some negative connotations that NGOs would rather not assume. Even more importantly, in the United States lobbying implies loss of 501(c)3 tax-exempt status with the IRS. Technically, charitable organizations are not allowed to lobby. If they want donations to their cause to remain tax deductible, groups must not lobby the members or staff of a legislative body. All of the groups in this study continue to hold this privileged status, except the Sierra Club, as will be explained in the beginning of chapter four. And the Sierra Club itself has the Sierra Club Foundation, which is a 501(c)3 entity. But in actuality, NGOs are allowed to engage in conservation action through administrative lobbying, public interest litigation, and public education. "Lobbying" in this sense allows for meeting and working with government agencies that implement relevant laws. Groups just cannot use their funds for specific legislative lobbying or electioneering activity. This analysis, thus, uses the term "lobbying" in its more limited sense.[119]

A second strategy, litigation, arose as a notable strategy among NGOs in the 1980s, particularly as obstacles emerged in the more traditional executive and legislative venues. Targeting the judicial branch instead of executive or legislative arenas, this practice widened the parameters of access for environmental groups. Of course, this strategy remains in its early stages of development, as the legal concept of "standing" was only expanded domestically in the United States in 1971. While even more obstacles exist for NGOs at the international level than at the domestic, the continued elevation of soft law provides real promise here, as seen in the work of the NGO Earth Island Institute (EII). EII demonstrated the most well-known application of this strategy in the 1990s by galvanizing the dolphin-tuna debate in United States District Court and then following through as enforcement of the Marine Mammal Protection Act (MMPA) quickly spilled over into the General Agreement on Tariffs and Trade (GATT) and subsequently the World Trade Organization (WTO). Ongoing litigation continues domestically in the United States, as groups such as Defenders challenge the Bush Administration's attempts to water down labeling requirements in the most recent MMPA reauthorization bill.

Three other strategies shown in table 3 are also considered mainstream but, unlike those discussed above, they are more data-focused. Scientific research, as well as technical consultation and collaboration, continues to emerge as a viable strategy for environmental NGOs. This is particularly true for those issues such as biodiversity loss and climate change where scientific grounding for policy decisions often remains ambiguous. Property acquisition and maintenance is a distinctive preservation strategy, often taking shape in the form of the debt-for-nature swaps conducted in Latin American states since the late 1980s. A fifth strategy, the administrative responsibility of monitoring actual agreements, also works within the context of the existing policies, specifically in the form of international environmental treaties such as CITES and the CBD. In these oversight roles, NGOs are critical links in ensuring compliance. They provide a much-needed oversight function for both negotiation and implementation.

That said, participatory strategies such as grassroots networking and education initiatives are conceptualized as the sixth and seventh possible strategies. Reflecting the popularity of civil society rhetoric in organizations at the beginning of the twenty-first century, participatory strategies are being utilized more and more. Some notable problems remain here, as will be discussed in chapters 3 and 4, including times when participatory rhetoric remains just that—rhetoric without any substantive action to follow it. Still, qualitative examples of application exist. Under such catchphrases as "community-based stewardship," environmental groups increasingly seek to "lobby" the public as well as governments themselves. Indeed, this book demonstrates that lobbying the public is a powerful strategy for NGOs, particularly when used in combination with mainstream approaches.

Theoretically, participatory strategies could also include an eighth and final strategic option, that of outright confrontation. Yet, due to NGO emphasis on "getting the word out" and sparking media coverage, these are best conceptualized for the purposes of this book as specific tactics within strategies six and seven above. First popularized by Greenpeace in the 1970s, confrontational tactics were followed in the 1980s and early 1990s by groups such as Earth First!, although Earth First! itself has since splintered. Tactics here range from passive resistance such as pickets and demonstrations to more drastic measures such as outright monkey-wrenching, popularized in Edward Abbey's famous 1975 novel.[120] The more radical tactics have lost favor in most environmental activist communities today, but passive resistance continues to be a common tactical ploy for many NGOs. In many respects, boycotts, pickets, and demonstrations are a direct attempt to extend grassroots networking and educational initiatives— and may be considered a subset of these participatory strategies. These seven main strategies outline how NGOs play a role in every stage of the policymaking process—from issue definition and agenda formation to treaty negotiation and actual implementation.

Supports and Constraints in Organizational Structure

As noted above, a range of NGO strategies exists from grassroots orientations to checkbook diplomacy. NGOs are equally diverse in terms of their organizational characteristics. Indeed, organizational characteristics often determine the specific strategies available to environmental NGOs. The *Yearbook of International Organizations* identifies seven aspects of organizational life as comparative indicators. These include aims, members, structure, officers, finance, relations with other organizations, and activities.[121] While these are applied simply for determining whether a group is an NGO, an adaptation of several of these indicators is fruitful for this analysis. General demographics, decision-making style, partnerships, targeted constituency, and strategic concentration are the five most significant characteristics of an NGO in terms of providing supports or serving as constraints with respect to transnational biodiversity protection.

General demographics, particularly as they shape the legitimacy and resiliency of an organization, should be noted. Age, expenditures, membership, and staff all deserve attention here, as they combine to form the expertise and financial capital that enhance the legitimacy of an NGO. Expertise has been well documented in the literature and refers to technical, scientific, or even administrative support. Financial capital is an obvious prerequisite to meeting basic operational objectives and, as such, an integral component to legitimacy. It should also be noted that transparency shapes legitimacy as well. The more transparent NGOs are better able to establish legitimacy in the minds of both access points and the general public. NGO characteristics that enhance legitimacy provide welcome support for the specific strategies that NGOs select in their efforts to protect transnational biodiversity.

Resiliency of an NGO is also demographically dependent. "Some NGOs fail to invest in development of administrative infrastructure and the institutional capacity building necessary for the long-term viability of their efforts."[122] Their expenditures, staff, and membership are too weak for such NGOs to continue as a viable force. On the other hand, the danger of becoming a bloated organization, one where organizational inertia obstructs substantive advance, now exists as a real threat in many large NGOs. Resiliency recognizes this and denotes longevity of an organization and ability to withstand changes in the international (and national) political climate. Organizational age combines with staff prestige and stability to improve resiliency and thus the ability of NGOs to effectively implement their strategies. NGO characteristics that enhance resiliency, then, also provide support for the specific strategies that NGOs select in their efforts to protect transnational biodiversity.

This book addresses decision-making style as a notable organizational characteristic. On one dimension, the decision-making style of NGOs ranges between hierarchical and decentralized. Hierarchical decision making refers to policy choices that are made by headquarters staff, without any local office input or electoral choice (i.e., referendums) from the membership. Decentralized decision making represents the other extreme in this continuum, where all decisions on strategies and tactics are made at the local level—and coordination among the various local NGO chapters does not exist. Both these extremes offer certain benefits, as will be discussed in chapter 4. On the whole, however, the more NGOs are able to achieve a balance between hierarchical and decentralized decision making, the more they provide support for the specific strategies they select in their efforts to protect transnational biodiversity.

On another dimension of decision-making style, this book considers transparency. This concept measures the openness of an organization, a particularly useful trait if one hopes to establish public acceptance. For instance, a breakdown on the percentage of NGO funds that go directly toward environmental campaigns (instead of administrative support services) enhances transparency when that information is available not only to those within an organization but also those in the general public.[123] This study thus defines transparency as access to the decision-making process, including budgetary allocations. Previous applications of transparency demonstrate its utility, especially in the context of improving treaty compliance.[124] Improving transparency of decision making provides support for the specific strategies that NGOs select in their efforts to protect transnational biodiversity.

A final dimension within decision making is imagination. Imagination measures the ability of an NGO to incorporate a multitude of perspectives in its decision-making process. Those with a healthy mix of perspectives are more effective than those with too few—or too many. Imagination is expected to be most useful to NGOs when moderately applied. Being too imaginative or too unimaginative will decrease effectiveness. The more NGOs are able to achieve a balance between imagination and stagnation, the more support their decision-making style provides for the specific strategies they select in their efforts to protect transnational biodiversity.

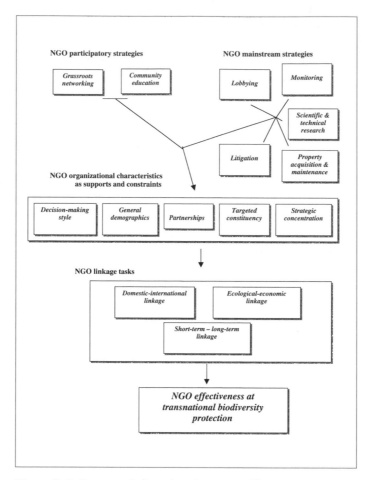

Figure 1.3. Framework for enhancing NGO effectiveness in transnational biodiversity protection.

In terms of organizational characteristics that act as supports and constraints, willingness to engage in partnerships is another area worth examination. NGOs clearly have enormous potential to influence one another by forming direct or even loose associations. These networks may be within a given state or across state boundaries. At times this is intentional. At other times it is not. In any case, alliances are important, as they augment NGO strategies with valuable support mechanisms.[125] Groups that neglect the potential synergies in establishing such networks unnecessarily handicap themselves. They limit the extent to which their expressed strategies can achieve any of the three fundamental linkages.

Targeted constituency is noteworthy. This relates to the previously discussed organizational distinction provided by Porter and Brown, that significant differences exist among national groups, international groups, and think tanks. This book supports the contention that only a transnational constituency can address adequately the issue of transnational biodiversity protection. For instance, while the *Yearbook of International Organization* treats membership as a dichotomous variable—with a group having dues paying members or not—a much deeper current exists. On one level, simple size of membership is important. On another level, the locations of these members, particularly in terms of nationality, are important. It is this dimension that is most critical in this analysis. The degree to which NGOs target their programs at a truly transnational constituency shapes effectiveness. Those that are transnational provide a valuable support mechanism for the specific strategies they emphasize in seeking to protect transnational biodiversity. Those that are not targeting a transnational constituency unintentionally constrain those same strategies.

Finally, strategic concentration is key. It takes into account the extent to which an NGO incorporates different strategies. Inflexible organizations are one-dimensional and inherently limited in this respect, whereas flexible organizations incorporate at least two different strategies. Of course, flexibility is most useful to NGOs when moderately applied. Being too flexible (just as being too rigid or inflexible) actually will decrease effectiveness. While a healthy mix of strategies can certainly enhance a particular initiative, NGOs can also go overboard and spread themselves too thin. Strategic concentration acts as a support mechanism for the specific strategies NGOs select when NGOs achieve a balance between flexibility and rigidity.

Framework Summary

Together, these organizational characteristics combine with mainstream and participatory strategies to create the three critical linkages required for effective biodiversity protection to take place. This path is depicted graphically in the flow chart in figure 1.3.

Working within the system: Mainstream strategies

2

INTRODUCTION

Nestled in northern Bohemia of the Czech Republic, in the infamous "Black Triangle" where the former Warsaw Pact states of East Germany, Poland, and Czechoslovakia met, lie the Jizera Mountains. The Jizera Mountains bore the brunt of the region's Soviet-era industrial engine for over forty years, a regime that showed little concern for the ecological base of economic development. Air pollution, acid rain, and unsustainable forestry decimated the region. Yet today, a remarkable recovery is taking place, including the enhancement of biodiversity with the return of native trout to local mountain streams. This trout reintroduction and watershed recovery is documented with over two decades of data collected by fifty-six-year-old Czech scientist Josef Krecek, a hydrologist from Czech Technical University. Krecek, who himself was not immune to the vagaries of Communist rule, struggled for years to publish his findings. Communist authorities stymied his professional advancement. They restricted his travel to the West. But ironically, today Krecek advises authorities in Prague, the Czech capitol where once his data was suppressed.[1]

Bringing further justice to past injury, Krecek was tapped in April 2003 by *Time Europe* as one of three environmental heroes in its inaugural list of those "who remind us what it means to make a difference."[2] Krecek is indeed an environmental hero. He was "the first to prove the link between industrial pollution and the dying forests and watersheds."[3] While working for Czechoslovak Forest Research Institute in the 1980s, Krecek found evidence that heavy industrial pollution was harming the area's forests and watersheds, watersheds that provided a fifth of the Czech Republic with drinking water. Indeed, the 32,000 square kilometer area of the Black Triangle was a graveyard for trees during the Communist regime. Coal-fired power plants pelted the forests with acid rain, damaging up to "two-thirds

Figure 2.1. Earthwatch volunteers monitor ecology in the Czech Republic. SOURCE: © Maureen O'Neill/Earthwatch Institute.

of the forested headwaters in the Czech Republic."[4] Krecek started studying the region in the early 1980s and continues to document the health of the region today as the primary investigator for the Earthwatch Institute-sponsored expedition, Mountain Waters of Bohemia.

Earthwatch began sponsoring his study of the region in 1991, and over a decade later remarkable progress is documented. Sulfur deposits have dropped by at least 20 percent, perhaps 50 percent in some areas, and lake and river systems region-wide are recovering. Biodiversity has benefited, particularly as the aforementioned native brook trout (brook char, *Salvelinus fontinalis*) have been successfully reintroduced thanks to Krecek's work in three mountain reservoirs. As Krecek and his research assistant, Zuzana Horicka, a doctoral candidate in hydrobiology from Charles University, explain in their letter to Earthwatch volunteers:

> The aim of our project is to contribute to a recovery of mountain eco-systems—first, to provide a long-term monitoring and scientific verification of the effects of acid rain and forest practices on the soil, plant and water interactions, and second, to support multi-resources land use practices, and to find a compromise between a specific land use and the air pollution control.[5]

The *Time Europe* award is not the only good news Krecek received in the last few years. Krecek served a two-year elected term as chair of the European Forestry Commission/FAO Working Party on the Management of Mountain Watersheds from 2000 to 2002. He also shared with Horicka the 2002 Scientist of the Year Award honor from Earthwatch. Indeed, Krecek's work under Earthwatch is a prime example of working within the system, demonstrating the role that the NGO mainstream strategy of providing scientific research can play in biodiversity protection. Interestingly, the Mountain Waters of Bohemia project also demonstrates the participatory strategy of environmental education. Volunteers who register for this Earthwatch Expedition perform a multitude of tasks, including evaluating Norway spruce populations, reintroducing brook trout, collecting water, soil, and vegetation samples, and taking pH, temperature, conductivity, and oxygen readings throughout some thirty-five different bodies of water. In short, these volunteers learn a good deal about the ecosystem and the biodiversity within it.

MAINSTREAM STRATEGIES

Let us take a closer look at precisely how the first dimension noted above plays out. Mainstream NGO strategies exist in essentially two forms. There are those that are political-action–focused and those that are data-focused. The political-action–focused strategies of lobbying and litigation garner the majority of popular press attention. This attention is much deserved, as NGOs have made significant strides inside the beltway and its power poli-

Table 4. Mainstream Strategies Utilized by NGOs Examined

	Lobbying	Litigation	Research	Property	Monitoring
BIONET	X		X		X
CI	X	X	X	X	
Defenders	X	X	X		
Earthjustice	X	X			X
Earthwatch			X		
Environmental Defense	X	X	X		X
Ocean Conservancy	X	X	X		X
Sierra Club	X	X	X		
TNC			X	X	
WRI			X		
WWF	X	X	X	X	X

tics circles.[6] One should note that, while NGOs seek to change the rules and adapt them to more favorable circumstances, these strategies still remain within the established process for inducing change. Targeting official state channels, they operate according to the standard methods of policy-making and remain mainstream—indeed too mainstream for the tastes of some. Those strategies that emphasize data collection or assessment—scientific and technical research, property acquisition and maintenance, and monitoring—constitute the other mainstream sub-grouping. As with the initial strategic initiatives of lobbying and litigation, each of these strategies is also categorized as mainstream because they operate within the established parameters of official state power.

Table 4 delineates which NGOs examined in this study utilize each of these five specific mainstream strategies, including those that are political-action focused and those that are data-focused. A cursory review of table 4 shows that scientific and technical research is the most popular option, while property maintenance is the least popular. On the other hand, the table does not address the relative degree of emphasis within an organization. For example, The Nature Conservancy emphasizes property maintenance and acquisition with scientific research driving this emphasis.[7] It should also be noted, as demonstrated in this table, that most groups utilize a variety of different strategies, and significant overlap does exist despite the five divisions above. This point notwithstanding, there are good examples where discrete application of one particular strategy is possible. Specific examples of each are addressed below.

Lobbying

As influential political scientist Harold Lasswell's famous dictum states, politics is the study of who gets what, when, and how. This distribution of resources, of course, depends fundamentally upon the concept of power and the art of persuasion that accompanies it. Of the multitude of options available, perhaps the most common tool in this endeavor is political lobbying. Individuals lobby. State governments lobby. Groups lobby. It is no surprise that lobbying emerges as the most popular choice of strategy for modern environmental organizations. United States history is replete with examples of lobbying by interest groups, and the contemporary currents of globalization magnify this significance in both kind and degree. When it comes to this particular strategy, despite their avowed nonparochial status and critical distinctions from interest groups in this regard, many environmental NGOs display characteristics remarkably similar to their interest group brethren. Yet, in very simple terms, NGOs are more diverse than interest groups, as the NGO rubric includes groups with interests that transcend national boundaries. NGOs are nongovernmental and as such not attached to specific political boundaries that traditionally limit interest groups. As the world becomes increasingly "smaller," though, distinctions between these two types of entity lessen as well. But one critical difference does remain. The NGOs in this study are all not-for-profit groups, including the Sierra Club, whereas many interest groups serve specific financial interests.

These NGOs still lobby or advocate, though. And a number of options exist for the twenty-first century, ranging from the electronically based mediums available just within the mid- to late 1990s; to the constituency-based approaches illustrated by various NIMBY battles popularized in the 1960s, 1970s, and 1980s; to the classical, direct lobbying applied throughout United States history.[8] Lobbyists typically target governmental representatives, either in the legislative or the executive branch—although NGOs must remember the restrictions against legislative lobbying or electioneering activity if they wish to retain 501(c)3 status with the IRS. NGOs thus often select secondary targets such as influential businesses and industries. Overseas lobbying efforts follow similar paths and NGOs expand this term in that, in addition to international institutions and states themselves, civil society itself becomes a primary target as these groups attempt to shape public opinion across the globe. This type of lobbying can be assigned the broader label of advocacy when it encompasses the more participatory strategies outlined in chapter 3, namely grassroots networking.

Groups such as Conservation International (CI) and Environmental Defense (formerly known as EDF) demonstrate the extent to which NGOs target governments, business, industry, and public opinion. One notable example combining the lobbying of state governments, international institutions, and public opinion is the ongoing effort to reform the World Bank and transform it into a more environmentally friendly institution. In the 1980s, efforts to publicize the destruction of the Amazon attracted widespread attention. Hollywood stars spoke out publicly, and environmental NGOs made concerted efforts to encourage policy reform in the World Bank. CI joined this fight, targeting several large-scale development projects that the Bank sponsored. Specific negative impacts from these projects included pollution and environmental degradation from mining as well as flooding from hydroelectric power. Road construction itself also threatened species diversity, since these roads both encroached on the previously undisturbed forests themselves and encouraged further migration to these ecologically fragile areas.[9]

With the United States powerful position courtesy of a weighted voting system within the Bank, CI targeted both the appropriations process in Congress and the upper levels of the Treasury Department, the executive agency responsible for communicating United States policy to its representatives at the World Bank. CI also went directly to Congress and testified at an array of hearings. With its premises rooted soundly in economic theory and incorporating the ecological-economic linkage, CI was effective in its lobbying efforts, and the Bank hired more environmental staff and finally integrated the costs of negative environmental impacts into their development decisions. CI is particularly proud of its role in creating a more transparent, accessible, democratic, and—most importantly they believe—an environmentally more friendly multilateral lending institution, according to Cyril Kormos, formerly of Conservation International.[10]

While many in the environmental community take issue with this last contention, it is clear that the Bank's Global Environment Facility (GEF) incorporated some of the recommendations that NGOs have advocated since the mid-1980s.[11] These small steps in a more environmentally conscious direction are noteworthy; the GEF, at $100 million a year, is the single largest source of funding for biodiversity protection. This is progress even if it only defrays the added costs countries incur when making environmentally beneficial adjustments to previously planned projects. World Bank President James Wolfensohn admits to being influenced by CI re-

garding conservation and publicly states that he "looks forward to continued efforts (with the World Bank and CI) to promote conservation and
sustainable development."[12] On the other hand, one cannot but wonder
how much of his statement is public relations posturing. Despite the efforts of NGOs such as CI, frustration continues to mount among those
dedicated to the fight to "green" the World Bank. Then again, some con-
servatives see GEF as going too far to the environmental left. Ironically,
they perceive it as a program created specifically to pacify environmentalist critics of the World Bank, one that has become a "giant protection
racket in which environmental activists shake down the WB in exchange
for their silence and even their support." In an op-ed attacking NGOs,
James Sheehan, director of international environmental policy at Compet-
itive Enterprise Institute (CEI), and Paul Georgia, an environmental research associate at CEI, express the fear that the American taxpayer is funding green pressure groups like WWF and Greenpeace, which are selected
by GEF to run such environmental projects.[13]

Litigation

Lobbying is not the only option available to NGOs engaging in political action through mainstream approaches. Groups are increasingly turning to
litigation as well. Domestically in the United States, litigation arose as a viable strategy when the legal concept of standing expanded in 1971 after a
local conservation group challenged an application by New York Edison
Company to build a power plant in the Hudson River Valley.[14] Prior to this
case, citizens could only bring suit when they themselves were affected by
pollution. An individual had the right to bring an issue before the court
only when personal injury or property damage could be clearly demonstrated. Needless to say, environmental groups found it difficult to qualify
as litigants under these conditions. Throughout the 1970s, though, this
precedent began to change, particularly as the 1969 National Environmental Policy Act (NEPA) spawned hundreds of lawsuits that challenged
agency decisions. For the first time, environmental groups were able to go
beyond cease-and-desist orders, and at times the rare fine, to actually sue
polluters when government officials failed to enforce the law.

Given the constraints of international law, though, opportunities to exploit the litigation end of mainstream approaches are not as great on the
international level as they are at the domestic level, although soft law, particularly as it evolves into customary law, presents real promise here. These
constraints aside, some degree of domestic implementation has obvious
implications internationally, as seen in the dolphin-tuna debate. Earth Is-

land Institute's (EII) litigation for dolphin-safe tuna in the United States engendered a debate in the early 1990s that gained notoriety in General Agreement on Tariffs and Trade (GATT) and subsequently World Trade Organization (WTO) discussions. In the late 1990s, the dolphin-tuna issue continued to spark international disagreement when the Ocean Conservancy actively sought a redefinition of what constituted dolphin-safe tuna. The MMPA had defined dolphin-safe tuna as tuna caught without the practice of "dolphin-setting," using the sighting of dolphins as the cue to drop huge tuna-catching nets. Dolphin and tuna swim together in the Eastern Tropical Pacific, apparently because they are feeding on the same type of food, although the science remains sketchy in regard to the precise relationship here.[15]

In any case, the Ocean Conservancy's rationale is that, for adequate protection to take place, international (namely developing-state) fishermen need incentives to meet the new regulations. Incorporating technological innovations might make this possible, but only if domestic considerations in the United States would adjust and recognize that developing-state tuna fleets could fish in a sustainable, dolphin-friendly manner. Working with Mexican and Latin American tuna fleets, the Ocean Conservancy has been instrumental in redefining the meaning of dolphin-safe tuna. By concentrating not only on United States government officials, but also on the fishing and science leaders in other countries, the Ocean Conservancy is striving to lay the groundwork for resolution of this highly contentious issue. Several humane organizations continue to oppose such a redefinition, though, and it remains a touchy issue for the lead proponents on both sides of this policy disagreement. Domestic pressures in the United States must recognize and accommodate domestic pressures in Mexico and other Eastern Tropical Pacific fleets—and vice versa—if a truly international agenda is to form.[16]

Perhaps the best illustration of litigation as a strategy is found in the efforts by Earthjustice, formerly known as Sierra Club Legal Defense Fund, an organization formed in 1971 in San Francisco (their Washington, D.C., office opened in 1978) by a group of attorneys volunteering for the Sierra Club. The mission of Earthjustice is to "bring the power of law and the judicial system into the fight to preserve the planet."[17] Clients total over five hundred and include fellow organizations such as the Ocean Conservancy and Defenders of Wildlife as well as the Sierra Club. Most of Earthjustice's work is domestically oriented around the endangered species act, its reauthorization, critical habitat designations, and forestry issues. Earthjustice does do some international work, though. Support of Colombia's

U'wa indigenous tribe against Occidental Petroleum is one such international example.[18] Earthjustice represented the U'wa before the Organization of American States Inter-American Commission on Human Rights, convincing the commission to hear firsthand the concerns of this group.

The U'wa are an indigenous group of about five thousand people who live in the high elevations of the Andean cloud forests in northeast Colombia. Believing that "oil is the blood of the earth," they threatened mass suicide in the late 1990s if Los Angeles–based Occidental Petroleum (OXY) drilled on their lands.[19] The threats to walk off a 1,400-foot cliff in mass suicide were rooted in a legend in which tribal members avoided Spanish enslavement by similarly jumping to their death. The modern-day threats, along with nonviolent civil disobedience, lawsuits, and letter-writing campaigns, gathered worldwide attention—and results. In late May 1998, OXY announced that it was leaving the territory in question, but was still seeking rights to drill on even larger U'wa ancestral lands. In response, according to the National Indigenous Organization of Colombia (ONIC), approximately thirty thousand Colombians went on strike and marched in demonstrations against oil development and in support of the U'wa from July to August 1998. Activism spread beyond Colombian borders, as the U'wa became a symbol of resistance to globalization and big oil worldwide. Environmental and human rights activists from Amazon Watch and Friends of the Earth to Project Underground and Rainforest Action Network have answered the call to support them.[20] Even then–Vice President and Democratic presidential candidate Al Gore was brought into the fray in January 2000 when protestors picketed his Manchester, New Hampshire, presidential campaign headquarters.[21] By July 2001, after a nine-year confrontation, the U'wa finally won their battle against Occidental Petroleum when OXY failed to find oil and began removing equipment from the site.[22] Still, the U'wa have no rest in sight, as the Spanish company Repsol recently began its own exploratory drilling on another tract of U'Wa ancestral territory.

Another notable foray into the international arena by Earthjustice is its loose partnership with a sister group in Canada, which still goes by the name Sierra Legal Defence Fund (Sierra Legal). Founded in December 1990, Sierra Legal Defence Fund provides "free legal services to environmental groups and concerned citizens" with its staff of over forty lawyers, scientists, and administrative support.[23] Today, Sierra Legal has offices in both Vancouver and Toronto and boasts a membership roll of over twenty thousand individuals. Earthjustice also holds alliances with a coalition of nonprofit environmental law organizations in Latin America. Groups in

Chile, Colombia, Costa Rica, Mexico, Peru, and the United States created the Interamerican Association for Environmental Defence (Asociación Interamericana para la Defensa del Ambiente) or AIDA in 1996.[24] Canada's Sierra Legal joined a year later in 1997. Through AIDA, this coalition of nonprofit legal groups bands together to help environmentalists throughout the Americas. It provides a forum for information exchange and the opportunity to counter multinational corporation threats in each of the seven states from which members come. Earthjustice also keeps abreast of trade agreements within the context of WTO issues. Earthjustice's 2002 docket included challenges to the WTO as it seeks to defend U.S. trade sanctions against states that fail to protect endangered sea turtles. Working with Fiscalia del Medeio Ambiente, Earth Island Institute in Chile, they seek to protect a United States law under challenge by India, Malaysia, Pakistan, and Thailand.[25]

Scientific and Technical Research

Many of these lobbying and litigation efforts are possible in large part due to another type of mainstream strategy—the use of scientific and technical research. While often counted among the least sexy of the strategies at the disposal of an organization, research has clearly emerged as one of the most popular—and the most effective. Research allows an NGO to define an issue, mobilize membership, access decision-making points, and even, at times, participate in actual implementation projects. Whether this research is purely scientific or merely technical support, it positions NGOs as an alternative to the more traditional information outlets, namely the states themselves. An example of technical research is found in Biodiversity Action Network (BIONET), an organization of various biodiversity-related NGOs that sought to strengthen overall NGO input into the Convention on Biological Diversity (CBD). With this mission, BIONET established an information clearinghouse in May 1996. This clearinghouse distributed information via the Internet to over a thousand leaders worldwide. Its guiding rationale was that stakeholders must first get on the same page in terms of the very information explaining an issue if they are ever to make any true progress on agreeing how to resolve that issue. BIONET provided detailed technical information as to how scientific findings apply in a given state's situation. It is important to note that this service is now only provided virtually on the Web, however, as BIONET ended its physical presence as a 501(c)3 in 2001.

Environmental Defense provides an example of an NGO undertaking avowedly scientifically oriented research.[26] From its humble beginnings in

1967 by a group of volunteer conservationists fighting to ban the spraying
of DDT on the marshes of Long Island, Environmental Defense now includes a staff of approximately 170 scientists, economists, attorneys, and
administrative assistants.[27] Born out of courtroom success thanks to the
merging of scientific and legal expertise, it has headquarters in New York
and other offices in Washington, D.C., Boston, Raleigh, Boulder, Austin,
and Oakland. With some three hundred thousand members and $41.2
million total operating expenses in 2002, Environmental Defense is large
enough to make its presence felt but is by no means as large as World
Wildlife Fund or The Nature Conservancy. It seeks to shape the legislative
agenda through lobbying and litigation. Yet, within the context of its original founding, Environmental Defense also seeks to bring science to bear
on public policy. Environmental Defense's famous campaign to convince
McDonald's to stop using foam plastic hamburger boxes is one example,
and within their International Program, Environmental Defense continues
to emphasize scientific research. By demonstrating that logging creates the
threat of flooding, soil erosion, and fire, the organization has halted public financial support of ecologically destructive projects in the Brazilian
Amazon, Indonesia, and central Africa.[28]

Property Acquisition and Maintenance

These types of scientific and technical research strategies, in turn, support
still another mainstream approach that works within the existing system.
Land acquisition and maintenance emerged as a unique strategy as early
as the 1950s domestically in the United States and became a viable international strategy in the mid- to late 1980s. While The Nature Conservancy
(TNC) engages in the scientific research discussed above, their most notable strategy remains property acquisition and maintenance.[29] Through
its self-anointed title as the real estate arm of the conservation movement,
TNC has carved out a respectable niche in most of the environmental community.[30] Incorporating in 1951 with an association of conservation biologists, TNC initially shared office space with the Wilderness Society in
Washington, D.C. Its first land acquisition was the 1955 purchase of a 60-
acre tract of bird habitat along the Mianus River Gorge in New York State.
TNC today owns over 116 million acres worldwide and supports programs
in the United States, Latin America, the Caribbean, Asia, and the Pacific.
They emphasize community-based partnerships and the pragmatic consideration of economics within an ecological context. TNC seeks market-
based results through non-confrontational benefits. An emerging popular
tactic is the debt-for-nature swaps that have multiplied throughout Latin

America in the past two decades. Thomas Lovejoy, then vice president for science at WWF, first suggested the debt-for-nature idea in an October 1984 op-ed in the *New York Times*. In a debt-for-nature swap, an NGO assumes a portion of debt for a third world state and converts that debt into local currency, which is then used for conservation purposes in that state.[31] While Conservation International (CI) instituted the first debt-for-nature swap, this practice has been adopted by various other organizations including The Nature Conservancy and World Wildlife Fund.

Monitoring

The fifth and final mainstream strategy, monitoring, again illustrates the high degree of overlap that exists among all five mainstream strategies. For example, the monitoring of biodiversity and the agreements designed to protect it clearly supports the first two mainstream options discussed, lobbying and litigation. While monitoring itself is obviously data-focused, it also provides valuable support for the mainstream approaches that are decidedly more action oriented. And some tactics of monitoring, such as the video footage that Greenpeace shoots of whalers on the hunt, undeniably, actively "bear witness." To a certain extent, monitoring of agreements acts as the NGO version of the United Nations blue helmets. And as within UN peacekeeping operations, lack of executive enforcement powers (and heavy physical weaponry) can be a decisive handicap for NGOs.

NGOs still are limited to the extent that they can participate in intergovernmental organizations such as the CBD. Often the level of participation depends on an NGO's relationship with a specific state. Some actually serve on state delegations, and, thus, have more rights than those that do not. Even if not given this level of access, the opportunities to at least observe are normally wide open, and NGOs are not precluded from having an impact in performing oversight functions.[32] Neutral oversight may have profound effects in not only law and policy development as a treaty is translated from negotiation to implementation stages. It also may impact the ultimate effectiveness of a treaty as flouters are reported publicly over the course of the existence of an agreement.

Established in 1993 at a meeting of NGOs in Washington, D.C., the aforementioned Biodiversity Action Network (BIONET) emphasized precisely this strategy. BIONET was formed as a direct response to the need for an NGO network on biodiversity issues. Membership did not in itself commit organizations to any specific policy position, since BIONET was simply a loose association of groups networking among themselves for the common purpose of enhancing biodiversity protection policies. That said, members did share a common mission to strengthen NGO input into the Convention on Biological Diversity process and to advocate effective implementation of the CBD worldwide. Specific objectives included the collection and distribution of information to other NGOs as well as IGOs, construction of a global network, strengthening of NGO input into the policy-making process, and support of efforts to achieve ratification and constructive interpretation in the United States and other states.

One way that BIONET met these objectives was through the strategy of monitoring. Its information clearinghouse, for instance, is a highly influential medium, providing open access to any individual or organization. Its listserv, one of only two or three globally devoted to this issue, included governments, NGOs, IGOs, and industry among its members. The original intent was to connect BIONET's clearinghouse mechanism with the official clearinghouse mechanism of the CBD Secretariat, but that linkage never materialized, as BIONET sought to retain its independence. BIONET believed that would provide more benefit than any official links would provide. This decision highlights a structural problem within the CBD's database, which is entirely governmentally owned. Countries provide content that is not peer-reviewed. Biases are not filtered out. With very little input from civil society in the official mechanism, then, groups must turn to the BIONET version for incorporation of native group perspectives along with the official governmental programs attempting to conserve biodiversity.

Summary and Findings

As we have seen from the discussion above, each of these five mainstream strategies contributes to the NGO objective of transnational biodiversity protection. Each also has its inherent weaknesses. Table 5 summarizes both of these.

Given these strengths and weaknesses, NGOs often apply these strategies in conjunction with one another as discussed above, and at times in tandem with the participatory strategies discussed in chapter 3. A plethora of combinations emerge here, with scientific research paired with property acquisition, monitoring paired with lobbying, litigation paired with lobbying and monitoring. One common theme is found throughout. Each of these strategies, or combinations of strategies, is more effective when framed within the context of the three fundamental linkages of domestic-international, ecological-economic, and short-term–long-term considerations. NGOs apply these strategies, both consciously and unconsciously, with these linkages in mind.

Table 5. Strengths and Weaknesses of Mainstream Strategies

	Strengths	Weaknesses
Lobbying	Draws attention to both the issue and the NGO itself as a viable political force	Still dependent upon sympathetic state sponsorship
Litigation	Adds another medium to pursue when legislative and executive channels are closed	NGOs still often shut out of legal proceedings at international level
Scientific and technical research	Provides hard evidence to NGO arguments and aids other strategic approaches in this regard	At times ambiguous with potentially competing scientific perspectives emerging to dilute a given argument
Property acquisition and maintenance	Ownership gives further legitimacy as stakeholder and allows NGO to dictate how a given tract of land is to be used	Does not address fully the causes of biodiversity loss, instead serving as simply a stopgap, short-term measure
Monitoring	Improves transparency of international agreements and aids those states that cannot afford to pursue oversight on their own	NGOs still dependent on states for approval to participate and NGOs hold no enforcement powers

FUNDAMENTAL LINKAGES

Domestic-International Linkages

The first linkage that NGOs are uniquely suited to develop is also the most difficult to achieve. Drawing heavily from the transnational theoretical literature outlined in chapter 1, the domestic-international linkage incorporates each of the three fundamental hurdles mentioned in the introductory chapter, most notably the conflict over power and self-interest as well as the lack of a global, ecological consciousness. A host of NGOs address this linkage, employing a litany of strategies. A common theme is that NGOs believe that they need to actually be in the regions where biodiversity is threatened if they are to link domestic and international considerations to one another. Rather than simply dictating policy out of Washington, D.C., environmental NGOs increasingly establish outposts, affiliates, or outright ownership in a given threatened area.

The Nature Conservancy provides one example by emphasizing the last point on ownership. The model for TNC from the very beginning was to

do conservation at actual sites. A staff of some 3,200 people in three hundred locations, including twenty-two different foreign offices from Indonesia and New Zealand to Japan and the Yunnan Republic in China to
Palau and the Federated States of Micronesia to Peru and Brazil, oversees
the largest private system of nature sanctuaries in the world. In Latin
America alone, TNC boasts relationships with forty-five other organizations in efforts to guide community development, professional training,
funding for legally protected areas, and debt-for-nature swaps. Philosophically, TNC believes that long-term impact is only possible if transnational
groups work directly with local organizations. It is truly a site-based organization whose method of saving land is simple. TNC buys it.

This approach has its share of hardships, as illustrated by the growing
pains TNC experienced when it began to sponsor property acquisition outside the United States. In attempting to make the domestic-international
linkage, TNC faced opposition to applying its standard United States approach in developing countries. In the context of large, rich, conservation
NGOs dictating policy from the outside, in retrospect, opposition should
not have come as a surprise. The perception, rightly or wrongly, was that
northern gringos were buying up the forest and having poor farmers ar-
rested if they wanted to cut firewood to cook their dinner that night. TNC
quickly came to the realization that their domestic model needed to be
modified for international applications. Operating according to a deeper,
not broader, slogan, TNC has gradually enlarged its international program,
although budget allocations remain weighted toward domestic operations.[33]

Supported in large part by the United States Agency for International
Development (USAID), TNC's Parks in Peril Program (PIP) serves as a good
example of property acquisition supporting the domestic-international linkage. Although some critics question whether programs at this micro level
can have an impact globally, the stated objective in this program is simply
to set a good example and influence the protection of other areas.[34] Since
1989, this effect has made its presence felt. TNC has spent $11.5 million
on this program, with USAID contributing another $42.5 million.[35] Some
sixty-five protected areas exist now thanks to PIP, as well as thirty-three
partner organizations in fifteen different Latin American states. While the
specific program components of sharing lessons learned, creating long-
term financing mechanisms, consolidating infrastructure, and integrating
local communities into the management of protected areas incorporate each
of the three fundamental linkages, the domestic-international linkage is
paramount. Bringing international concerns to the actual domestic location, and inversely domestic concerns to the attention of international ac-

Figure 2.2. Parque Nacional del Este on Saona Island in the Dominican Republic. SOURCE: © Connie Gelb/Courtesy The Nature Conservancy.

tors like TNC, makes effective biodiversity protection possible. Parque del Este, pictured in figure 2.2, is part of TNC's Parks in Peril Program.

Located on the southeastern tip of the Dominican Republic, Parque Nacional del Este is one of the Caribbean's largest marine parks. Four of the world's seven sea turtle species nest along the shores here as well as eight species of birds found nowhere else but on Hispnaiola, the island the Dominican Republic shares with Haiti. Over-fishing of snapper, grouper, conch, and lobster as well as the unsustainable hunting of the Caribbean monk seal threaten this region. TNC's partners linked up with the Dominican Department of Fisheries and used the PIP data to develop a law establishing protective zones and a finite fishing season for conch nursery grounds in Parque Nacional del Este. All told, Parks in Peril has protected more than 28 million acres in the Caribbean and Latin America. These initial successes helped TNC, with renewed USAID support, launch a $30 million initiative for Parks in Peril in October 2001, expanding it to twelve additional areas and providing management training in hundreds of other sites.

Another such example is found in Guaraquecaba, Brazil, where the largest remaining tract of Atlantic forest has been conserved in two private reserves, thanks to the mainstream strategy of scientific research (and the participatory strategy of community education) targeting the domestic-international linkage. TNC recognizes that, in the final analysis, their Parks in Peril program will only succeed if supported by the local people. It does not matter how many pieces of paper are signed if this condition is not met. More and more, TNC emphasizes the needs of local communities in its efforts to protect biodiversity.[36] To be effective, TNC must link the domestic and international on this strategic approach. More than those who finance land acquisition must deem it of importance. The people who actually live there must both subscribe to this belief and have incentives to continue to support this subscription if the land is to be protected.

Looking to the Ocean Conservancy, a relatively small international NGO but still the largest marine biodiversity organization in the world, provides another example of an organization incorporating the domestic-international linkage. Despite emerging relatively recently on the international environmental scene, the Ocean Conservancy has made a significant impact in environmental NGO circles. Since its inception in 1972, other, larger environmental groups have recognized the previously neglected issues of marine species diversity and formed programs devoted to marine wildlife within their respective organizations. Today, virtually every major environmental group has a marine program of some sort. Bob Irvin, former vice president of marine wildlife conservation for the Ocean Conservancy, foresees continued growth in this area internationally as the Ocean Conservancy and other NGOs expand the number of areas in which they work.

The Ocean Conservancy's species-specific scientific research under the auspices of the World Conservation Union and its efforts to develop greater scientific understanding of general ecological relationships are an integral component of their strategic agenda. But it is in lobbying states on the Convention on International Trade in Endangered Species (CITES) that the Ocean Conservancy has made some of its most visible inroads toward establishing a link between domestic and international concerns. In their position as an accredited observer, the Ocean Conservancy has participated in virtually all of the Conference of Parties (COPs) to CITES. One should note that lobbying here supports this domestic-international linkage in one of two distinct respects. As Reshma Prakash, senior correspondent at *The Earth Times,* points out, NGOs both support and challenge states at the international level.[37] According to her experiences covering various conventions on the environment and sustainable development, NGOs are increasingly important in that they at times serve as delegates themselves, most notably with marine issues such as whaling. At other times, NGOs play just as important a role—but from the other side of the fence. NGOs definitely target delinquent states, Prakash states, providing balance when it is needed. The Ocean Conservancy utilizes both of these options, at times assisting Latin American states in their attempts to meet dolphin-safe tuna requirements and at other times pushing African states to abide by ivory trade regulations under CITES.

Indeed, lobbying is a favorite strategy of organizations such as the Ocean Conservancy, although lobbying is commonly referred to as "advocacy" within the community, perhaps because "lobbying" carries a more negative connotation and draws obvious parallels to business interests but also due to the aforementioned IRS restrictions. In any case, lobbying organizations such as the Ocean Conservancy do not deny the power of official, governmental status. From their experience, being an official member of a UN organization carries more weight than being an NGO precisely because it allows them to link domestic and international concerns. Staff members from time to time even serve as official United States delegation members to the Northwest Atlantic Fisheries Organization. That is, the Ocean Conservancy actually has a seat at the table, and thus, more influence on the United States position. Despite occasional opportunities of this nature, though, the vast majority of the impact windows NGOs encounter are from the outside looking in.

A good example of this is found in another initiative at the Ocean Conservancy, a scientific and legal exchange program with Cuba begun in late 1999. Combining science and law is not a new phenomenon in the environmental community, as we have seen most notably with the origins of Environmental Defense, although application of this combination in Cuba certainly is. Bob Irvin, former vice president and general counsel at the Ocean Conservancy, was part of this effort to draft environmental protection laws in Cuba. The Ocean Conservancy received every required license from the Treasury Department, "carefully dotting all their i's and crossing all their t's." In fact, Irvin states that the Ocean Conservancy received tacit encouragement from the U.S. government.[38] Government officials seem to hope that this initiative could help establish a solid working relationship for future endeavors. With an eye to the future, spillover is expected to occur into other areas beyond the environment. Ironically, though, politics had nothing to do with the Ocean Conservancy's initial decision. Science did. The Ocean Conservancy believed that Caribbean protection efforts required working with Cuba. As Irvin states, the organization first asks itself "what does the science tell us we should be doing" when it tackles any new issue. Virtually all their work proceeds from this question.[39] It is science-based advocacy.

But politics cannot be divorced from this objective, as the Ocean Conservancy discovers from time to time. Their biggest problems are not with the biannual renewal process. What have been most difficult are the occasional Cold War mind games that still take place even at the close of 2003. While David Guggenheim, vice president for conservation policy, carefully couches his criticism of this outdated Cold War thinking, making a point to commend both the U.S. government and Cuba for allowing the Ocean Conservancy's various projects to develop over the last ten years, he also admits to exasperation with periodic political flare-ups. For example, an important component of the Cuban program is the aforementioned exchange initiative that brings Cuban scientists to the United States to attend conferences. "That is help they desperately need," notes Guggenheim. "It helps build good will. But visas are sometimes approved the day after the conference starts. [It's] a Cold War tit-for-tat and caught in the middle are hard working scientists trying to save the planet."[40]

The Ocean Conservancy also engages in a number of multilateral programs. One example of linking domestic to international considerations multilaterally is found in their annual International Coastal Cleanup campaign. Begun in 1986, this program recruits enthusiastic volunteers from around the world to pick up trash along the coastal waters of their state. While the strategy of participatory networking is implicit within this strategy, the Ocean Conservancy's Coastal Cleanup campaign also incorporates the mainstream strategy of monitoring. Data collected from this cleanup have been used to determine sources of pollution and thus serve as valuable information in monitoring existing agreements. This type of coastal cleanup program also complements what Irvin believes will be increasingly emphasized within NGO circles, the need to actually be in the places where the biodiversity is located rather than simply operating out of Washington, D.C. Working on areas that complement each other and finding the synergies between various initiatives, Irvin conceptualizes a future effectiveness that is largely dependent upon the degree to which this fundamental linkage is achieved. Whether it is dolphins and tuna, endangered sea turtles, or sharks, the Ocean Conservancy integrates this philosophy in its three key mainstream strategies of lobbying, scientific research, and litigation.

The domestic-international linkage is also incorporated into the strategic thinking of Conservation International. CI, like many of the large, nationally based, environmental NGOs, utilizes a variety of strategic approaches to complete its overall mission. In addition to political and economic lobbying, CI uses combinations of scientific research, litigation, and community involvement. CI was founded in January 1987 in the lounge of the Tabard Inn, a quaint Dupont Circle hotel just five blocks from the White

Figure 2.3. Ocean Conservancy volunteers help in International Coastal Cleanup. Volunteers pick up trash in Baltimore, Maryland, in September 2002. The International Coastal Cleanup is held annually the third Saturday in September. People in one hundred countries, including all fifty-five of the United States and territories participated in 2002, gathering over 8.2 million pounds of trash. SOURCE: Photo coutesy of The Ocean Conservancy. "Over 8.2 Million Pounds of Trash Collected During 2002 International Coastal Cleanup," May 5, 2003 Press Release.

House. Within six months of its founding, in July 1987, CI sponsored the very first debt-for-nature swap. Debt-for-nature swaps, such as this initial one in Bolivia, allow developing state governments to cancel debts by agreeing to mobilize domestic resources for environmental initiatives, although this amount is typically miniscule when considered in the context of the state's overall debt. Another caveat here is the fact that the resources from this debt reduction that are then allocated to environmental funding are also relatively small compared to the need that exists. These points aside, debt-for-nature swaps are still helpful in promoting environmental awareness, strengthening NGOs, and even protecting small pockets of biodiversity.[41]

Today, CI boasts more than a thousand staff working in more than thirty countries on four continents but is perhaps best known for its "hotspots" program that targets highly threatened, high-biodiversity areas. Begun in 1990, the program has been compelling to a number of donors, with the MacArthur Foundation an early supporter. Throughout the 1990s, the hotspots program continued to be a tool for CI to wield in grant competitions. Then in October 2001, CI received the largest grant ever for an environmental NGO. Totaling up to $261.2 million over a period of ten years, this grant came from the newly established Gordon and Betty Moore Foundation based in San Francisco. It was given expressly to target hotspots and tropical wilderness by beefing up the Global Conservation Fund and establishing Centers for Biodiversity Conservation in Brazil and the Guianas, Madagascar, the Andes, and Melanesia.[42] As Keith Alger, vice president of the conservation strategy department asserts, "Hotspots are the places where species are likely to go extinct first, so that's where we want to spend money first—on the hotspots."[43] Combining the strategies of scientific research, lobbying, and property acquisition and maintenance, CI conducts biological analyses to determine what species are present in given areas and then distributes this information through its Rapid Assessment Program (RAP), launched in 1990. After making short-, medium-, and long-term viability assessments, CI implements specific initiatives designed to foster species protection.[44] This has drawn prominent attention. *National Geographic Magazine* launched a two-year series on biodiversity hotspots, beginning in January 2002 with a focus on India's Western Ghats. The map in figure 2.4 shows these hotspots.

Pushing to get the last pristine areas of the world set aside before it is too late, CI actively supports the creation of national parks in conjunction with local communities in its country programs. This approach attempts to link various domestic constituencies in a particular region, but, as CI's Cyril Kormos asserts, further ties are needed between what is going on in D.C. and what is going on in the field. High-speed expansion at CI over the last few years must not neglect this need for even closer ties between the various country programs and Washington offices. While the need for such ties obviously brings out the significance of grassroots involvement, Kormos believes that NGOs need more than just this participatory activism. They need to take that next step, he says, contending that participatory activities are only a preliminary step, that a progression from participatory to mainstream strategies advances the potential for policy-making impact. Chapter 3 takes issue with this proposed sequence of strategies, contending that both mainstream and participatory strategies are more effective when used in conjunction with one another. Still, Kormos makes a strong point to the extent that layers of activism buttress one another. Characterizing CI as more cautious about what it says and more reluctant to get into large-scale campaigns, Kormos thinks the strength of his organization lies in the technical analyses as well as legal and biological expertise that it pro-

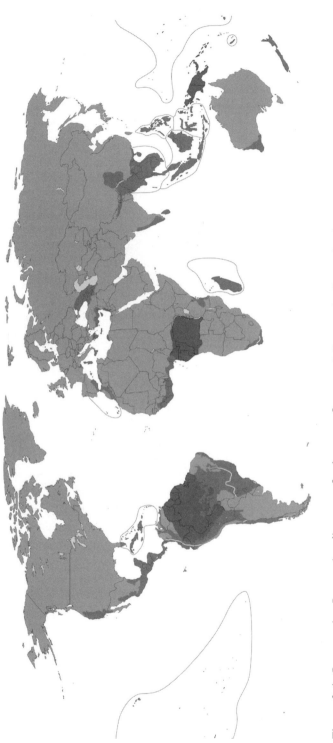

Figure 2.4. Conservation International's twenty-five hotspots. SOURCE: © Conservation International.

vides to the environmental community. As Kormos asserts, "We do a better job of doing our homework . . . providing a more accurate presentation of the landscape. We add layers of activism that others cannot. We are much bigger, with more resources."[45]

Kormos raises an interesting question in regard to size and financial purses. With a Washington staff of 150 to 160 people and another 450 individuals working for CI worldwide, does CI have a higher probability of being effective than smaller, less financially endowed organizations? Chapter 4 explores this issue in greater depth. But it should be noted here that, while money and size help, they fail to guarantee effectiveness. Many other factors are at work. The sophistication of CI's work allows it to back up threats, as Kormos argues, allowing CI to influence the World Bank and their appropriations, as discussed previously. Yet, even the largest and richest organizations are limited in their ability to influence if they fail to meet the domestic-international linkage.

Nels Johnson, deputy director of the Biological Resources Program at World Resources Institute (WRI), for one, raises the point that large domestic NGOs cannot be effective on a case-by-case basis in biodiversity protection without incorporating international public support. Despite high-profile public images, effectiveness in linking the domestic and international requires more than a steady cash flow. It relies heavily on the linkages between economic-ecological and short-term–long-term concerns as well as the extent to which the NGO remains transparent to the general public. Localized impact programs implicitly target broader audiences. The transparency-like contingencies that Johnson lists for the local level are actually key components of the domestic-international linkage as well. To go beyond the micro level, organizations must employ precisely this type of thinking. While partnerships and associations certainly enhance the likelihood that an organization will make the domestic-international linkage requirement, any NGO seeking to make an international policy impact, not just influencing localized biodiversity protection initiatives, must incorporate some degree of this linkage in its specific strategies.[46]

At times this linkage is not so obvious. One case in point is found in the work of Defenders of Wildlife, an ostensibly United States–oriented NGO. Yellowstone National Park, the Southwest, and Alaska attract much of the attention at Defenders. Through congressional lobbying, for example, Defenders seeks to preserve and strengthen the United States Endangered Species Act, specifically habitat for the gray wolf. Following successful reintroduction of gray wolves into Yellowstone in 1995, Defenders turned to the court system and various litigation efforts as the most promising venue to protect this policy victory. This fight has an international, or at least regional angle as well, though. Wolves of the northwest often skip customs, failing to recognize our border with Canada. Defenders recognizes the significance of this shared border and is currently mobilizing Canadians against the Northwest Territories winter wolf hunt with a mix of lobbying and litigation as well as more participatory events. Through a partnership with Earthjustice Legal Defense Fund, for instance, Defenders filed a petition to impose trade sanctions on Canada for failing to protect trans-boundary endangered species.[47]

Biodiversity protection is complicated by the fact that, in the grand scheme, individual efforts made at one locale will only be effective if they are multiplied at other areas throughout the world. Regional examples of success must be replicated for the issue of transnational biodiversity protection to be managed effectively. This is a contribution that groups such as World Wildlife Fund make. As WWF president Kathryn Fuller explains, "But what separates WWF, I think, is our ability to think and act globally— to translate decades of on-the-ground conservation experience into action at national, regional, and even global scales."[48]

That is, what makes the scientific research, lobbying, educational initiatives, and grassroots networking that WWF utilizes truly effective is the fact that they are applied throughout the globe. WWF actively pursues the domestic-international linkage. It recognizes that an array of local communities must join forces if any meaningful impacts are to be achieved. This requires recognition that no single group can achieve all this alone, that help is needed. It also requires recognition that various domestic constituencies have different agendas, agendas that often conflict with one another.

Such disagreement is best illustrated in the rift between developed and developing states over economic development and environmental protection. In a sense, the domestic-international divide here is really more accurately stated as a domestic-domestic (or state-state) division. Domestic constituencies in the states of the North such as the United States emphasize environmental protection, although they also stress open access to genetic resources as the natural heritage of humankind. Domestic constituencies in the South stress economic development—and the right to "benefit sharing" of the genetic uses Northern corporations have found for various third world natural resources. NGOs are the best instruments to make this linkage. To be effective, they must link these divergent domestic constituencies to create an international constituency. Such an international constituency incorporates the needs of developing states without compromising their future—and for that matter the future of the developed states as well.

This is a tall order, and to date NGOs have not been effective in meeting the domestic-international linkage. Most clearly, neither the second nor the fourth indicators of effectiveness, mobilization and implementation respectively, have been met. Environmental NGOs have either failed, or in some cases not even attempted, to bridge different domestic constituencies and form an international one. This is a fundamental component of mobilization of membership and staff, let alone the general public. Implementation, similarly, is conspicuously absent in the domestic-international linkage. This is best explained by the fact that, while some degree of access to decision makers does exist for NGOs, fundamental problems regarding issue definition remain. Too often, the issue of transnational biodiversity protection remains defined in state or, at the most, regional terms. Despite ecological warnings to the contrary, states continue to see the issue in pure realist state power terms. With this in mind, let us now turn to a discussion of the ecological-economic linkages.

Ecological-Economic Linkages

The second critical linkage required for effective transnational biodiversity protection is the connection between ecological and economic considerations. As with our previous linkage of domestic-international considerations, NGOs are uniquely suited to make this linkage, and all five mainstream strategies play an important function here, albeit one that encounters its share of obstacles as well. Indeed, as discussed in chapter 1, the dilemma of sustainable development complicates effective negotiation of the second requisite linkage. "Sustainability" has accumulated different meanings, depending on the particular perspective and interests of a given entity. In many respects, this is the very difficulty discussed above in the attempts to establish a domestic-international linkage. Domestic interests often hold ecological and economic interests in a position that is at odds with various international interests. Many environmental NGOs now believe that the domestic-international divide can only be successfully negotiated after the second linkage is completed. The evidence here supports that contention, that linking ecological and economic considerations serves as a precursor to linking domestic and international considerations. In that sense, the fact that the linkage of ecological-economic considerations is more fully developed than our first linkage bodes well for future environmental NGO efforts. Bridging the divide between ecological and economic considerations may be seen as the first step toward bridging the divide between domestic and international considerations. Before examining this suggestion further, however, let us first examine the ecological-economic linkage itself.

Bruce Rich, international director at Environmental Defense, is but one of many individuals who believe any route to solving contemporary global environmental problems must begin by "following the money."[49] Strategies must be formed around this fundamental concept, Rich asserts, and environmental NGOs may perform an integral function once this flow is identified. Through their lobbying initiatives, NGOs strive to divert development funds to more environmentally benign purposes. They act as the levers of change toward more sustainable development. Unfortunately, when and where to place these levers remains a contentious issue—within economic circles of government and multilateral lending institutions like the World Bank as well as within the environmental community itself. As noted earlier, the dolphin-tuna debate surrounding the amendments and enforcement of the 1972 Marine Mammal Protection Act is a case in point. Eventually most environmental groups, led by the Ocean Conservancy, Environmental Defense, Greenpeace, and National Wildlife Federation, signed onto the position that the MMPA needed to redefine its consideration of dolphin-safe tuna to incorporate technological advances that provide economic incentives for developing country fishing fleets. Some groups such as Defenders and several humane society organizations, though, continue to oppose any adjustments in the definition. To them, these proposals leave open the very real possibility that protection will be watered down in future labeling.

Along these lines, several organizations were leery about the November/December 1999 World Trade Organization (WTO) talks in Seattle, fearing that a schism comparable to that which occurred during the 1993 NAFTA debate was possible. Defenders of Wildlife, for one, made every effort to keep its agenda within the framework of its fellow environmental organizations. Although operating with only a small international staff, Defenders saw the trade talks as an opportunity to shape environmental policy at the international level—and desperately wanted to present a unified front. When these talks broke down and were postponed in the wake of huge protests from both environmental and labor camps as well as a fair number of violent anarchists, Defenders' fears failed to materialize. Of course, the premature closure of the Seattle round also precluded any chance to present a unified front. As seen in the essentially ineffective front presented by

NGOs at the April 2000 meetings of the World Bank in Washington, D.C., moreover, environmental NGO coalitions aimed at influencing international trade agreements remain in their early stages of development.

One way to build consensus for key international negotiations like WTO is to develop extended working relationships on the hill. This makes even more sense considering the assessment by Nicholas Lapham, former biodiversity protection officer at United Nations Foundation. As Lapham states, congressional appropriations represents the arena where the real action is.[50] It is this work on the hill and in the trenches where the nitty-gritty economic nature of ecological issues is truly exposed. Unfortunately, this is not as sexy as other approaches and, as such, is one place where a lot more work could be done. World Wildlife Fund (WWF) is one organization that seeks to fill this void. Staff members there devote substantial time and energy to lobbying appropriations in Congress, particularly in terms of foreign aid and assistance bills. Funding of USAID is a prime example and one that exhibits direct linkage of economic and ecological considerations. Franklin Moore of the Environment Center within USAID's Global Bureau reiterates this point. Moore also believes that environmental NGOs are most effective when they partner with local development NGOs in a specific region. By linking economics to their ecological agenda, these NGOs end up fostering democracy and governance that, in turn, further supports effective biodiversity protection.[51]

An increasingly popular and productive avenue in establishing the ecological-economic linkage is to target nongovernmental sources such as major multinational corporations. Environmental Defense, TNC, and CI each employ this strategy. Corporations formerly labeled as the evil enemy are now sought out as partners by several major environmental NGOs. This form of lobbying corporations, popularly known as "greening" business, underscores a division within the environmental community on a number of fronts. To many skeptics, for instance, the active association of an environmental group with a specific business or industry implicitly condones the production and consumption that a company sponsors. More liberal groups question whether these efforts are in fact positive contributions in the long term.

Sharon Beder, head of the Science, Technology and Society Program at the University of Wollongong, Australia, cites WWF, TNC, Defenders, NRDC, EDF, Audubon, and NWF as accepting company support from those who also sponsor anti-environment groups.[52] This can be dangerous. As Beder asserts, the "work with" industry mantra of TNC "can undermine environmental protection in the long term by 'greenwashing' these organizations and enabling them to continue with other damaging activities out of the spotlight of public scrutiny."[53] A series of front-page articles in the *Washington Post* in May 2003 echoed these concerns. These articles recognized TNC as perhaps "the leading proponent of a brand of environmentalism that promotes compromise between conservation and corporate America."[54] *Post* reporters David Ottaway and Joe Stephens went on to take TNC to task for a series of "costly misadventures and awkward positions." They allege that a "bucks and acres" strategy of raising money to buy land has compromised TNC's integrity on issues such as drilling in the Arctic National Wildlife Refuge and global warming, as a number of corporations on TNC's board are active in the camp that discounts the dangers of both these ecological threats. The irreverent national radio commentator Jim Hightower agrees with this general assessment. He believes that there is a real danger in getting too close to business and chastises mainstream organizations for playing nice, arguing that NGOs must instead get down and dirty if they really want to make progress. As only Hightower can, he contends:

> We've simply got to get the hogs out of the creek. As Aunt Eula knew, this is not a chore to undertake in your best trousers, politely pleading: 'Here hog, here hog . . . pretty please.' To get hogs out of the creek, you have to put your shoulders to them — and shove . . . Yet most national environmental organizations today are indeed dressed in their Sunday trousers, engaged in the soft-hands work of lawyers and lobbyists in Washington, sincerely but futilely attempting to negotiate the relative positions of hogs.[55]

TNC disagrees. Unlike most other large environmental NGOs, it is not really a policy advocacy organization. TNC takes positions on some issues but only if they meet the criteria of being conservation related and boast diverse support in a distinctly bipartisan manner. TNC also makes a conscious effort to take the high road and to avoid any negative battles. Instead, TNC praises people for positive contributions, never denouncing groups or individuals no matter how negative their impact—although TNC does not deny the need for this type of approach within the environmental movement at large. The direct result of this approach is that a broad assortment of people form alliances with TNC, including some who would never make a "green" list in even the most generous of assessments. Again, this is one reason why more vocal environmental groups criticize

TNC. But perhaps there is also a degree of jealousy as well, for through their checkbook diplomacy, TNC has indeed built the largest private sanctuary system in the world.

Shifting gears somewhat, the production and consumption of sun-grown coffee, which began in the early 1970s, has prompted another initiative where some degree of economic success is present, albeit on a much smaller scale than TNC's land acquisitions program. But the potential is there. A number of NGOs have taken steps to capitalize on the fact that, after oil, coffee is the most widely traded legal commodity in the world, and that the shift to sun-grown coffee has had a devastating impact on species diversity. It is with this in mind that Defenders, for one, selects coffee grown in areas that are not only known for their taste but are also rich in biodiversity. Defenders' Java Forest is grown by small family farms and boasts both organic and fair-trade certified designation, meaning that the farmers that grow the beans get a fair price for their product.[56] Java Forest is grown in the mountains of Oaxaca in southern Mexico and the western highlands of Guatemala, with plans to expand in the near future to Nicaragua and Peru. As mentioned earlier, Conservation International also has its own program, one that began with its August 1999 shade-grown coffee initiative with Starbucks. It struck a deal for financial and technical assistance to the tune of a $150,000 grant over three years for small, shade-growing coffee farmers in the Chiapas region of southern Mexico.[57] Starbucks sold out of this blend in short order. Aside from the fact that growing coffee in the shade means that one does not need to cut down trees, shade-grown coffee also happens to be much less dependent upon fertilizers and pesticides than sun-grown coffee. And fertilizers and pesticides, in turn, present obvious hazards to fragile ecosystems and the threatened species that inhabit them. This is of real importance to biodiversity protection, as there is often significant overlap between where biodiversity is and where the coffee-growing hotspots are, as Glenn Prickett, CI business and policy vice president, asserts.[58] Privately, however, even individuals who oversee these programs are quick to caution that the long-term contributions of this initiative remain in doubt. Starbucks will likely continue the relationship, but some 50 percent of the coffee in the world is now produced by the ecologically destructive, sun-grown method.

Scientific research, too, has economic implications worth mentioning, albeit in a slightly more indirect manner. Earthwatch Institute, based outside of Boston in Maynard, Massachusetts, serves as a good example here. Earthwatch has three listed fundamental objectives: conservation, research, and experiential education. In reality, though, their concentration and strategic niche is the synergy achieved by linking research and education. They utilize thousands of volunteers annually to gather data in the field for the various research projects they support throughout the world. The express purpose of using volunteers in this research is to help raise the level of public awareness. Earthwatch not only wants to practice solid scientific research, it wants to use the act of conducting that scientific research to encourage personal responsibility, to get people (through the volunteers and perhaps the messages they take home to friends and family) to grasp the economic relationships to the ecological areas in which they serve. Earthwatch hopes to create more responsible consumers. It hopes to help create and maintain the emerging global civil society—and ensure that this civil society understands what a sustainable future really entails. Earthwatch is interested in the domestic-international linkage, but the organization recognizes that this linkage will only be possible if the ecological-economic linkage is met first. Once people understand the connections between their consumption decisions and the environmental ramifications of that consumption, then the domestic-international linkage is feasible. Until then, Earthwatch believes energy expended expressly in that direction is futile.

Still another example of an organization utilizing economic incentives as a new approach to solving environmental problems is found in Environmental Defense. *The Wall Street Journal* considers Environmental Defense "one of the hottest environmental groups around."[59] With no prospects for the United States becoming a party to the CBD any time soon, Environmental Defense has directed a good deal of its attention to biodiversity within the Endangered Species Act here in the United States and to aspects of biodiversity internationally as it relates to the Climate Convention. The group used to actively participate in CITES, but that has increasingly become a low priority for them. With groups like WWF making huge investments in monitoring for CITES, Michael Bean explains that there is no need to be just a voice in the chorus in this case, poignantly noting, "Why bang your head against that wall. We have better things to do with our head."[60] In a round-about away, Environmental Defense targets international biodiversity in various sustainability programs, particularly those attempting to shape wood products selection. For instance, Environmental Defense's Steve Schwartzman, a leading authority on Brazilian indigenous communities and environmental issues in the Amazon, often brought Chico Mendes, the famous Brazilian rubber tapper and grassroots activist, to speak with World Bank officials. Mendes' presence at Environmental Defense meetings enabled people from within local communities to make their

Figure 2.5. Chico Mendes in front of the Xapuri Worker's Union holding his son Sandino in early 1988. Mendes was killed in December 1988 by cattle ranchers who opposed his rubber tappers union, but his spirit lives on as a true environmental hero. Courtesy Barbara Bramble, Director, Alianza para la Vida Silvestre, National Wildlife Federation.

case internationally, clearly an act in support of the domestic-international linkage. Serving as a facilitator in this manner, Environmental Defense also put a human face on the issue and drew direct ecological and social connections from the economic decisions of the Bank.

Today the motto of Environmental Defense could be paraphrased as simply: whatever works. When it was founded, the organization regularly invoked the simple strategy of "sue the bastards." But they soon found that litigation seldom solved an issue in a definitive way, and began to supplement their litigation skills with other tools. Tom Turner, senior editor at Earthjustice, agrees with this assessment. In his mind, "Litigation is a powerful strategy, but again it has to work in concert with everything else." Turner also cautions, that "litigation is never the last word . . . but it can buy you time."[61] Increasingly at Environmental Defense, and noteworthy for the distinct strategic niche it has carved out, this litigation pairs with lobbying and research to attach economic significance to conservation. That is, Environmental Defense expressly targets our second linkage. The Alliance for Environmental Innovation, based in Boston, is a good example. This program grew out of the Paper Task Force and its idea to encourage recycling by recruiting corporate partners. While admittedly several steps removed from direct biodiversity protection, the extent to which Project Alliance succeeds means less pressure for forest products here in the United States, which in turn puts less pressure on destruction of habitat essential to biodiversity internationally. This type of initiative also points to the need to draw connections between short- and long-term objectives as they relate to biodiversity protection.

The lobbying strategy of CI supports this linkage as well. As seen in the congressional appropriations for USAID, the World Bank, and regional development banks for Africa, Asia, and Latin America, both bilateral and multilateral projects increasingly include specific CI-recommended directives toward biodiversity conservation.[62] Similarly, Defenders employs lobbying with the express intention of linking ecological and economic considerations. For instance, they actively opposed legislation that would have given the president "fast-track" authority to negotiate new trade agreements, fearing that such agreements would fail to secure environmental protection. Using the lobbying strategy, Defenders sought to preserve existing ecological-economic links and to make sure that any future trade agreements incorporated this approach.

Lobbying can also take the form of advocacy at international meetings. WWF and its affiliates often engage in this strategy. At times, the specific purpose of this is to draw a direct correlation between ecological and economic considerations. The release of the WWF report, "The Impact of Dams on Life in Rivers," on the eve of the World Commission on Dams in April 2000, in Cape Town, South Africa, is a good example. Compiled from data on ninety-one dams in thirty different countries, the report shows that some 250 species are threatened by dam construction. By releasing this report just before the World Commission on Dams convened, WWF attempted to redefine the context of the meetings, to push state governments toward better management of their river basins and wetlands.[63]

Scientific and technical research, as well as expertise, is another strategy that contributes to the ecological-economic linkage.[64] CI's Forest People's Fund, which supports a range of development activities in southern Suriname, including distribution of benefits derived from Bristol-Myers Squibb and their 788 extracts prepared from 394 plants, is one good example. The Global Biodiversity Forum, which is convened regularly in conjunction with the Subsidiary Body on Scientific, Technical and Technological Advice to the Convention on Biological Diversity, is another. In a series of workshops, numerous NGOs, including the Ivory Coast affiliate of WWF, World Resources Institute, and BIONET, offer both general and specific technical advice to state delegations on implementation of the CBD. In three separate workshops held in 1999, participants discussed how to integrate biodiversity into regional plans and programs, how to use ecosystem approaches in managing the biodiversity of dry lands, and how to better adapt management considering issues of scale. More specifically, recommendations for integrating biodiversity into tourism, agriculture, forests, and fisheries were outlined.[65]

All told, these mainstream strategies have been remarkably effective to date in establishing the ecological-economic linkage. The ability to mobilize staff, membership, and even the general public has been crucial here. Access to decision makers who directly influence ecological-economic connections also plays a prominent role. But the degree of effectiveness is perhaps most noteworthy to the extent that actual implementation of the ecological-economic linkage takes place. While these initiatives admittedly are on the local level thus far, they do expressly incorporate ecological and economic considerations. One could also argue that they have laid the requisite groundwork for initiatives that are more global in scale. To make this final jump, however, environmental NGOs first will have to negotiate one final indicator of effectiveness, that of issue definition itself. Indeed, it is this first indicator, ironically, that has proven the most problematic when it comes to the ecological-economic linkage, primarily in terms of the ambiguous concept of sustainable development. Once this condition is met, once parties can come to terms on what sustainable development really looks like, then the initial steps that environmental NGOs have made toward bridging the divide between ecological and economic concerns can be fully cemented. While some degree of divergence will always be present, ecological and economic desires clearly need not be submitted to endless zero-sum conflict. States can incorporate both in their development agendas. Continuing to develop this approach, however, will require integration of the third and final linkage, negotiating conflicting demands between short-term and long-term considerations.

Short-Term–Long-Term Linkages

NGOs are uniquely suited to make the linkage between short-term and long-term considerations. More so than states, in fact, NGOs incorporate long-term objectives in their specific program initiatives. This is a critical component in protecting transnational biodiversity, as seen in the ecological portion of the previous chapter. Only when policy-making decisions, both within states and within the larger international arena, are made from a multi-generational perspective can species diversity be truly protected. Several examples bear out this contention, illustrating both its significance and the shortcomings NGOs exhibit to date on this critical linkage.

While BIONET engaged in other strategies such as research and lobbying, and contributed to the domestic-international and ecological-economic linkages, it is within its monitoring strategy and the contributions of this strategy to the short-term–long-term linkage that its most significant impact may be found. Granted, monitoring in this case clearly supports the linkages between different domestic constituencies and assists in the creation of an international constituency. Stakeholders must get on the same page in terms of information before true progress can be made. The CBD itself has an official clearinghouse mechanism, but states do not have an independent authority to which they must answer. BIONET, on the other hand, was open to environmental, developmental, and any other citizen-based NGO in the United States. While the governing steering committee floated the idea of expanding targeted states beyond the United States in the late 1990s, expansion on that order never did occur before the NGO ceased physical operations in 2001.[66] That is, technically, only United States–based NGOs were members.

BIONET sought to serve as the "eyes and ears" of the United States environmental community at the global level. With its foundation as a direct outgrowth of the concern among environmental groups that the United States needs to be a party to the CBD, it is easy to see the implicit domestic-international linkage within its agenda. With a fairly broad membership that shifted little over the years, BIONET was continually frustrated by United States intransigence on this treaty. Recognizing this, BIONET attempted to maneuver itself into position to fill the short- and long-term linkage void. Much of what the environmental community does involves dealing with short-term crises, as former BIONET coordinator Hans Verolme observes.[67] The legislative process itself is a prime example. It also illustrates why BIONET had to reorient its purpose. Given United States intransigence, BIONET shifted its orientation to implementation of the CBD. Looking first at marine biodiversity, then at forests, BIONET provided substantial technical research contributions to the COPs and their agenda focus on various ecosystems.[68] By most accounts, however, these agendas failed to develop beyond their genesis stages. As Verolme states, they just did not seem to go anywhere. Negotiations on forests largely fell victim to the fact that forests mean much more to people than biodiversity and, as such, much more than the CBD was at work. Despite these cold, hard facts, the environmental community continued to perceive forests as merely ecosystems during the COP discussions of the CBD. With these obstacles, only a fact-gathering, weakened compromise emerged.

These attempts represent BIONET's efforts to be forward looking, to link short- and long-term considerations. With only a virtual existence today, a new group must now take its place. Then again, perhaps one reason BIONET did not last was that it tackled this difficult linkage, one that had no rewards politically. As Verolme states, BIONET conceptualized in a time horizon of five years. Despite distinctions between his organization and those who lobby and litigate on the Hill in timeframes of two years and less, he believes these divergent organizational strategies still can be mutually supportive. "I cannot do my work if I am not comfortable with the work that my colleagues do in lobbying Congress. I cannot do my international work without those who do work in the U.S. on the ground here . . . and vice versa," he contends. It is here that BIONET provided a particular task for the environmental movement as a whole. As Verolme states, the little bit of work that BIONET did fits into a much, much larger picture. BIONET extended the time horizons of traditional policymaking, even if only for a few years.

One more example of a specific BIONET project that targeted the short-term–long-term linkage was its document proposing strengthening of the Global Environment Facility (GEF). Along with Birdlife International and IUCN, BIONET coordinated the production of this joint NGO document in March 1998. Points made in the study include the hope to infuse the GEF with a genuine learning culture, to integrate global environmental concerns into the non-GEF operations of its three implementing agencies themselves (the World Bank, UNEP, and UNDP), to expand the number of GEF implementing agencies, to enhance the leveraging of GEF by developing better coordination and synergies among funding institutions, and to outline the role of NGOs with particular attention to the concerns of developing country NGOs.[69]

The Ocean Conservancy is another example of a group that uses monitoring to support the short-term–long-term linkage. With financial support from the United States Environmental Protection Agency, the Ocean Conservancy monitors marine debris through its National Marine Debris Monitoring Program (NMDMP). Like the International Coastal Cleanup campaign discussed earlier, this program evaluates both the quantity and content of debris—although the NMDMP is limited to only U.S. coastlines. The Ocean Conservancy then shares this information with the U.S. EPA, National Marine Fisheries Service, National Park Service, and U.S. Coast

Guard so that the United States government can develop policies to combat marine pollution.[70] The expressed objective is to link current pollution problems with long-term threats that marine pollution present to fish harvests and general ocean health.

Conservation International also performs a monitoring function that addresses the connections between short-term actions and long-term objectives, although unlike BIONET and the Ocean Conservancy, the monitoring that CI performs focuses upon its own work. As it explains on its Web site, CI wants "to know if [our] activities achieve the conservation objectives we set for ourselves within the biodiversity hotspots. We want to assure our donors that their investment truly conserves biodiversity." Their Monitoring and Evaluation program "is a tool to plan for the next cycle."[71] Implicit in this language is the short-term–long-term linkage. CI chooses individual species of animals and plants to act as indicators of conservation success. Their Ecosystem Monitoring Network (EMN) provides a forum for information exchange by connecting staff and affiliate organizations. Of course, this network also fosters the domestic-international linkage, as experts from throughout the globe are able to interact with one another and with officials at CI. Still, the overriding concern is to connect short and long-term demands. As with the BIONET examples above, CI also faces the obstacles of state sovereignty here. States remain reluctant to make short-term sacrifices when long-term gains are not easily quantifiable. During the Mediterranean Plan debate, for instance, Algerian president Houari Boumedienne succinctly stated the standard opposition platform to long-term environmental gains. According to President Boumedienne, "If improving the environment means less bread for the Algerians, then I am against it."[72]

These same constraints are seen in the obstacles that confront the World Wildlife Fund (WWF), particularly its department of government relations. As one of the truly global environmental NGOs, WWF has its hand in virtually every mainstream strategy outlined earlier in this chapter. Lobbying bills the old-fashioned way, WWF engages in a series of briefings and congressional maneuverings regarding foreign operations appropriations. USAID funding is one such example, as Estraleta Fitzhugh, a congressional liaison in WWF's department of government relations, explains.[73] Development assistance bills that increase funding to USAID cannot be effective unless they have an environmental component. This direct application of the linkage of ecological-economic considerations deserves mention in this section, because it is also expressly linking short-term economics and long-term economics in both the United States and the specific developing state that receives the USAID assistance. By attending meetings throughout the year, testifying before committees, and proposing draft language for actual legislation, Fitzhugh seeks to make just this link.

Such approaches, unfortunately, are limited by the fact that lobbying is typically a two-year process. Most of the time, getting what you want takes at least two Congressional sessions, according to Fitzhugh. And even then, a certain amount of luck never hurts. Being in the right place at the right time, a combination of luck and persistence, allows groups like WWF to emerge victorious in the traditional "slushing it out" process of policy making within the United States. By supporting capacity- building or democracy governance, long-term agendas may be incorporated within the short-term actions of the legislative process.[74] It is only in this realm that one finds real opportunities to transcend short-term dominated agendas. An excellent example is the connection of environmental issues to health issues such as infectious diseases.

Litigation also has an important role in achieving the short-term–long-term linkage. For one, simply by virtue of going to court, an NGO such as Earthjustice Legal Defense Fund raises community awareness of an issue. Even if they lose in court, this exposure may be enough to create victories in the future. At times, environmental groups pursue litigation with precisely this agenda in mind. For instance, once community awareness is raised, similar projects that threaten biodiversity may never even get off the ground. Even if they do, they will likely be less environmentally destructive than they would have otherwise. Quite simply, as Friends of the Earth contends, "a court case that fails may still attract grassroots attention that can lead to new laws."[75] This litigation generally comes about in one of two ways: either it is part of a concerted strategy to achieve a specific reform or it is the last resort of an NGO that has been stymied in lobbying or other efforts. What is also interesting is the tendency for environmental NGOs to build partnerships when they engage in the strategy of litigation.[76] Group efforts with other environmental groups and at times with social justice organizations not only enhance the position of an NGO on that particular lawsuit, but these partnerships also tend to build lasting relationships that buttress long-term agendas. They pave the way for future success. They explicitly target the short-term–long-term linkage.

To shape long-term agendas, though, NGOs must first be accepted as legitimate stakeholders. One strategy applied with this goal in mind is the previously discussed property acquisition employed by TNC. While this clearly ties ecological and economic considerations, as outlined in the previous linkage section, it also supports the short-term–long-term linkage. As Terry Shultz, the Conservancy's stewardship ecologist in Colorado states, "Ownership also gives you a seat at the table so you can see the direction a community wants to go. You're not usually asked if you're perceived as an outsider."[77] Negotiating the short-term–long-term linkage, as with the other two linkages, requires an expansion of recognized stakeholders on an issue. Only then will the appropriate conditions exist to protect transnational biodiversity adequately.

NGOs have made notable inroads toward achieving the third linkage of connecting short-term and long-term considerations, thanks to the five mainstream strategies discussed above. At least on the local level, NGOs display a remarkable ability to meet the demands of access and implementation, the third and fourth indicators for effectiveness outlined in the first chapter. On the other hand, NGOs have been unable to mobilize sufficiently either their staffs or memberships, let alone the general public, when it comes to connecting short-term actions to their long-term ramifications. The first indicator of effectiveness, issue definition, helps explain this deficiency, for NGOs also have been unable to define adequately the issue of transnational biodiversity protection within the context of the short- and long-term linkage. It is within this context that future effectiveness will depend, not only upon meeting all four criteria of issue definition; mobilization of staff, membership, or general public; access to decision-making points; and actual implementation of projects. Future effectiveness will also rely on the ability of NGOs to join each of the three fundamental linkages to each other.

Summary and Findings

Joining the three types of linkages is not an impossible task. As seen in the discussion of linkages, considerable overlap already exists. Programs that incorporate the domestic-international linkage often address the ecological-economic linkage. Initiatives that emphasize short- and long-term connections integrate ecological and economic connections. Sometimes this overlap is unintentional. Yet, many other times, this is by design, as NGOs recognize that the complexity of the issue of biodiversity protection demands exploitation of the dependencies within these relationships. This chapter demonstrates the gaps in each of the three linkages as addressed by mainstream strategies, most notably that in the domestic-international arena. It also demonstrates the contributions that specific mainstream strategies make toward the three linkages, particularly in the ecological-economic arena. Table 6 summarizes those groups that utilize mainstream strategies in establishing the three fundamental linkages discussed in the preceding pages, including a rough distinction in intention and actual result.

Several patterns stand out in the schematic in table 6. First, it is clear that the majority of strategies are devoted to the first two linkage hypotheses; domestic-international linkages and ecological-economic linkages. Second, there is a stark distinction between intent and actual impact, a point that will be addressed shortly. NGOs, no matter what specific strategy they employ, display a remarkable record in terms of establishing the ecological-economic linkage. Whether they use lobbying, litigation, scientific and technical research, or property acquisition and maintenance, NGOs make a solid case in this linkage. Unfortunately, the same cannot be said for their success with the domestic-international linkage. While there are actually more cases of NGOs utilizing mainstream strategies in this linkage, overall effectiveness has yet to be achieved in this arena. Granted, specific initiative examples of success do exist. There is also remarkable potential to translate these micro cases to the macro level. However, to date, that type of leap has not been made, and some in the environmental community question whether it ever will be.

One final point should be made about the schematic before we turn to the specific issue of effectiveness. The column representing the linkage of short- and long-term considerations is relatively sparse compared to the other two. This is a reflection, I would argue, of policy making more generally as a short-term endeavor. Politicians and policymakers rarely think beyond the immediate issues. Environmental problems domestically were plagued with this deficiency for much of the twentieth century. Now in the twenty-first century, as we grapple with environmental issues on an even more global scale, these same sorts of obstacles confront foreign policymakers. Environmental groups must recognize that the daunting timeframe within which this task is posed does not preclude it from consideration. In fact, this urgency dictates its need. Environmental NGOs must not fall victim to the policymakers' trap of fighting only the short-term battles. Particularly when addressing issues such as biodiversity protection, only a long-term perspective—albeit with concerted links to immediate, short-term concerns—will have the chance to protect diversity adequately for future generations.

Effectiveness here, as well as with the other two linkage hypotheses, relies on the ability of NGOs to meet each of the four criteria outlined in the latter half of chapter 1 and further discussed at the end of each linkage section in this chapter. Table 7 summarizes the status of the effectiveness of

Table 6. NGOs' Mainstream Strategic Contributions to Three Linkages

	Domestic-international linkage	Ecological-economic linkage	Short-term–long-term linkage
Lobbying	BIONET CI Defenders Earthjustice Environmental Defense Ocean Conservancy Sierra Club WWF	**CI** **Environmental Defense** **WWF**	BIONET WWF
Litigation	CI Defenders Earthjustice Ocean Conservancy WWF	**Ocean Conservancy** **CI** **Environmental Defense** **Sierra Club**	Earthjustice
Scientific and technical research	BIONET CI Defenders Earthwatch Environmental Defense Ocean Conservancy TNC Sierra Club WRI WWF	**CI** **Earthwatch** **Environmental Defense** **TNC** **Ocean Conservancy** **Sierra Club** **WRI** **WWF**	BIONET Ocean Conservancy WRI
Property acquisition and maintenance	CI TNC	CI TNC WWF	TNC
Monitoring			BIONET CI Ocean Conservancy

Note: The regular type above reflects the NGOs' intended connection between the three main linkages and its strategic objectives. Boldface type indicates when a particular NGO actually meets these strategic objectives, thus fulfilling the linkage requirement.

Table 7. Effectiveness of Mainstream Strategies in Establishing Three Linkages

Domestic-international linkage	Ecological-economic linkage	Short-term–long-term linkage
Some degree of access present, but issue definition remains problematic and neither mobilization nor implementation really exists to date.	Good measure of success with mobilization, access, and even implementation, but ambiguity of definition itself (i.e., sustainable development hurdles) continues to frustrate full completion of this linkage.	Able to access decision-making points and implement projects on local level, *but* still struggle with definition and mobilization.

mainstream NGO strategies towards establishing the three fundamental linkage hypotheses. Several patterns stand out in the schematic. First, as noted earlier, the domestic-international linkage is woefully inadequate. Neither mobilization nor implementation exists in this linkage, and significant problems remain in terms of the issue definition itself—that biodiversity threats in one state or region are biodiversity threats for all states. Similarly, while the degree of access that NGOs have to decision-making points improved significantly in the 1990s, states and their IGOs continue to hold the upper hand. NGOs, within the current international framework, will always be secondary actors, dependent upon the good graces of state sponsors.

Second, in regard to the ecological-economic linkage, NGOs display a higher order of effectiveness. Mobilization of membership and, in some instances, the public at large, clearly exists in numerous regional instances. Access to decision-making points continues to aid the linkage of ecological and economic considerations. Even implementation itself often incorporates this linkage, at least on local levels. Despite essentially meeting each of these three conditions, though, NGOs continue to find their effectiveness limited by the initial indicator of issue definition, namely the ambiguity that surrounds the term "sustainable development." Questions remain as to whether (and how much) Northern states should help foot the bill for ecologically friendly adjustments to Southern state development. Questions also remain as to whether Northern states should share the benefits derived from various natural resources that their multinational corporations identify in those Southern Hemisphere states. Until these issues are resolved, full completion of the ecological-economic linkage is not possible.

Third and finally, the short-term–long-term linkage also exhibits a mixture of effectiveness and ineffectiveness. Access and implementation indicators are often met, as we have seen, but issue definition and mobilization remain problematic. On a continuum of effectiveness for these three fundamental linkages, then, the short-term–long-term linkage falls between the first two. More indicators for effectiveness are met than in the domestic-international linkage, but not as many as are met in the ecological-economic linkage. This pattern suggests some practical policy adjustments for environmental NGOs in the field, at least those that focus upon mainstream strategies. If a progression of steps is at work here, NGOs should first tackle continued enhancement of the ecological-economic linkage, namely negotiating the controversies that surround issue definition. This would not only enhance the ability of NGOs to meet this fundamental linkage; it would also make the short-term–long-term linkage more feasible, and, in turn, increase the probability of achieving the domestic-international linkage.

Integration of the various mainstream strategies described in the chapter is a key step in constructing a support structure that facilitates precisely this perspective. While no single approach can negotiate this road alone, each of the five mainstream strategies makes significant contributions to the three fundamental linkages between domestic and international considerations, ecological and economic considerations, and short- and long-term considerations. The next chapter demonstrates that the mainstream strategies above are not enough, that participatory strategies of grassroots networking and community education initiatives are needed as well. The central point within this chapter, though, is that, while mainstream strategies may not be sufficient in and of themselves in protecting transnational biodiversity, they are necessary components in the overall equation. That is, each of these strategies provides a crucial function in species protection efforts.

Admittedly, varying degrees of overlap exist here. At times, for example, there is virtually no distinction between technical research and monitoring. Similarly, lobbying and litigation often go hand in hand in many of the larger, national environmental organizations of today. This type of overlap initially drove this analysis, as my expectations were that the really interesting work, and not coincidentally the really effective work, would be that which combined these different strategic dimensions, particularly in combinations of mainstream and participatory initiatives. At least when considering solely mainstream components, this chapter demonstrates that such logic is only partly right. On the one hand, no organization effectively engages all five of the mainstream strategic dimensions. Such attempts are too taxing on the resources of an organization.

If an organization tries to do everything, to perform each necessary component of the sufficient conditions for effectiveness, it will undoubtedly end up doing nothing very well. Attempting such an unadvisable task not only dilutes the effectiveness of an organization on a particular issue; it also damages the long-term viability of the organization overall. A similar argument can be made for those organizations that seek to branch out to new issue areas. Many in the environmental community feared in late 1999, for example, that BIONET was doing just this in taking up consideration of bio-safety and reorienting itself toward environmental trade agreements. Within two years, the organization was defunct, at least in physical terms. NGOs that assemble their strategies in an ad hoc fashion lose valuable credibility points in the policy-making community and significantly weaken their ability to implement individual programs. On the other hand, as indicated by the discussion above, certain initiatives are effective precisely because they employ imaginative mixes of the strategic dimensions available.

Working with people: Participatory strategies

3

INTRODUCTION

Sometimes a picture is worth a thousand words. Throw in a couple catchy headlines and a series of full-page advertisements in *The New York Times* and they are priceless, enough to save the Grand Canyon. That is precisely what the Sierra Club did in June 1966 when the first of four full-page ads ran in an all-out effort to raise public ire against the United States Department of the Interior plans to dam the Grand Canyon. With headlines like "Who Can Save Grand Canyon? You Can . . . and Secretary Udall can too, if he will," as well as "Now Only You Can Save Grand Canyon From Being Flooded . . . For Profit," and "Should We Also Flood the Sistine Chapel So Tourists Can Get Nearer the Ceiling?" the Sierra Club won a remarkable battle—and signaled a fundamental shift in its mission. The Sierra Club had been flirting for years with its tax-exempt status for donations; internal documents show that it was investigated four times by the Internal Revenue Service even well before the turbulent tenure of David Brower.[1] But the advertisement shown in figure 3.1 pushed them over the edge, with an IRS representative hand-delivering a letter the day after its publication that signaled that a tax status change was imminent.

Brower, who is known in environmental circles as the archdruid of environmentalists, was appointed as the first executive director of the Sierra Club in 1952 and stayed in that post until he resigned in May 1969.[2] He cannot take sole credit for those 1966 advertisements, as advertising agency executive Jerry Mander penned the most effective lines in terms of the number of responses returned to the Sierra Club, and Ansel Adams likely provided the Sistine Chapel rhetoric in a personal letter to Brower shortly before it ran near the end of July 1966.[3] But certainly under Brower's leadership the Sierra Club shifted away from the "paternalistic philanthropy" tradition of the past to a future firmly rooted in grassroots support.[4] He

Figure 3.1. Grand Canyon battle advertisement. The first of the Grand Canyon battle ads appeared on June 9, 1966, and sent shockwaves through the IRS, Washington, the Sierra Club—and America. SOURCE: Roderick Nash, *Grand Canyon of the Living Colorado* (New York: Sierra Club/Ballantine Books, 1970), 132. Michael M. Gunter, Jr., thanks the Colby Memorial Library, Sierra Club for use of its historical archives.

modernized the Sierra Club. Membership skyrocketed in the three years after the advertisements ran, doubling to seventy-eight thousand by June 1969. And while the Sierra Club did lose tax-exempt status for its donations, the Grand Canyon energized this new membership in ways that are still felt almost forty years later.

PARTICIPATORY STRATEGIES

Sierra Club's use of a newspaper campaign is also notable in that it bypasses traditional structures. It is participatory. It emphasizes working with people. Despite being a domestic example and not directly targeting biodiversity, this is a precisely what the above story about the Sierra Club's "battle ads" illustrates. They asked Americans to mail brief notes to their government. They encouraged citizens to take action. This is crucial. While mainstream strategies lay a solid foundation for NGOs seeking to protect transnational biodiversity, they often fall short in completing the three pivotal linkages that form the cornerstone for truly effective protection efforts. Full connection between domestic and international, ecological and economic, and short-and long-term concerns requires participatory approaches as well as mainstream initiatives. Participatory approaches directly involve those who have a stake in an issue. They target the people themselves with the objective of getting those people to commit to a cause, to act. Participatory approaches, thus, enable NGOs to better take advantage of their unique situation and exploit the transnational nature of environmental issues generally as well as biodiversity protection specifically.

Those groups that employ participatory approaches fill a much-needed vacuum in the community by personalizing the story that is being told. While mainstream strategies incorporate some of these tactics, with advertising campaigns on television and through the Internet, their energy is focused upon official government channels. Participatory approaches, on the other hand, deal with civil society. They address the direct costs and benefits that specific communities incur, a task that more mainstream approaches often find daunting given the constraints of the traditional policy-making process. Quite simply, participatory approaches tap directly into human nature and make the issue of transnational biodiversity protection more tangible than if only mainstream strategies were utilized. Participatory strategies personalize an issue within the minds of their membership and the public at large.

This is an important contribution. Drawing connections directly between people and their environment, both immediate and distant, fosters individual action. It makes people want to act. The participatory strategies of NGOs facilitate correction of past deficiencies in environmental policy-making circles, namely the erroneous conceptualization of various mediums as distinct and impermeable. Only with the birth of the modern environmental movement, particularly its roots in Rachel Carson's *Silent Spring* and the public controversy surrounding misuse of chemical pesticides in the early 1960s, have policymakers begun to take into consideration the environmental connections among different policy decisions, especially as they affect human health and well-being. NGOs are uniquely suited to take advantage of these relationships both domestically and internationally. Nowhere is this more evident than when NGOs apply the strategic approaches that employ participatory democratic theory. Bringing citizens actively into the biodiversity protection process enhances understanding of the fundamental linkages discussed above and adds much-needed power to the entities attempting to inspire change.

Participatory strategies span the entire political spectrum. At times, certain initiatives even include both ends of the spectrum at the same time, bringing together groups with widely divergent interests. The not-in-my-backyard (NIMBY) phenomenon is noteworthy as an example where previously antagonistic entities may join together in their opposition to a common threat. Yet participatory strategies may provide the most potential in terms of what they propose to advance—not for what they oppose. For instance, with transnational biodiversity protection, mainstream approaches make their most significant contributions by opposing actions that threaten species loss and wildlife habitat destruction. Granted, buying a property, as well as some lobbying, litigation, and research, can aggressively advocate a particular policy of protection. When mainstream strategies are used on the offensive in this manner, they should not be couched in reactionary terms. Still, the overwhelming majority of mainstream strategies fall instead in the defensive category, manifestations of a reaction to an event or proposal. Even the strategy of property acquisition and maintenance may be categorized as one of exclusion, and, thus, one that is defensive.

This is not a criticism. Such approaches are a necessary component of virtually any policy-making strategy. Defensive maneuvering and well-conceived opposition tactics are essential. But often times they are not sufficient, particularly in an issue such as transnational biodiversity protection. There must be a good cop to the mainstream strategies bad cop. Someone needs to explain the benefits of an action, not just the negative consequences to inaction. Participatory strategies provide this by drawing direct connections to how the average person benefits from these policies.[5] Participatory strategies, as such, are largely offensive strategies couched in a more positive light than their mainstream brethren. They allow the envi-

Table 8. Participatory Strategies Utilized by NGOs

	Grassroots networking	Community education
BIONET		X
CI	X	X
Defenders	X	X
Earthjustice		
Earthwatch	X	X
Environmental Defense	X	X
Ocean Conservancy	X	X
Sierra Club	X	X
TNC	X	X
WRI		X
WWF	X	X

ronmental movement to move beyond the negativity manifested in a fight against enemies that plunder the planet. In its stead, environmental NGOs can adopt decidedly more positive rhetoric by framing their activism as a fight for the people who depend upon a healthy and viable biological diversity.[6] Whereas mainstream strategies stress reaction, then, participatory strategies stress action.

This chapter examines the role of two such participatory strategies employed in efforts to protect biodiversity; grassroots networking and community education initiatives. Table 8 delineates which NGOs utilize each of these two participatory strategies. As demonstrated in this table, most environmental NGOs that employ participatory strategies utilize both grassroots networking and community education initiatives. Only Biodiversity Action Network and World Resources Institute apply community education without grassroots networking. Earthjustice essentially applies neither. This trend is not surprising, since grassroots networking requires a degree of community education to be effective. This overlap also points to the remarkable potential of combining different strategic initiatives for the benefit of each linkage objective.[7] Before addressing these synergies, however, let us first examine the specific contributions of these two strategies individually.

Grassroots Networking

The Sierra Club, which holds the distinction of being the oldest grassroots environmental group in the United States, serves as a prime example of grassroots networking. Founded in 1892, it boasts a membership of over

550,000 and an annual budget of $42 million. Much of the Sierra Club's agenda targets domestic political issues—and even a few relatively apolitical issues that emphasize recreational enjoyment. One example is the series of domestic outdoors events and hikes that local Sierra Club chapters hold for their members. Over the last several years, headquarters adopted similar programs, expanding to reach beyond United States borders.[8] A September 2000 trip to South Africa celebrating its magnificent diversity is one case in point.[9] Where the Sierra Club truly makes its contribution to the larger environmental NGO community, though, is in the arena of developing public participation in the decision-making process. More specifically, with its emphasis upon the strategy of grassroots networking, the Sierra Club finds its strategic niche.

Biodiversity protection, while still an important component of the Sierra Club's international agenda, receives consideration largely as a derivative of other program initiatives. These related programs include global warming, responsible trade, human rights, and stabilizing world population.[10] Interestingly, though, one could make the argument that the relationship between these programs and biodiversity underscores the extent to which biodiversity protection is so important, that all other programs relate to it. On its global warming Web site, for instance, the Sierra Club states, "the effects of global warming may be measured in extinctions, not degrees."[11] The postcard shown in figure 3.2 underscores that connection. This postcard was self-addressed to then–Vice President Gore and encouraged tighter automobile pollution regulations, obviously for clean air purposes but also more generally to curb global warming and thus counter species die-offs.[12]

While the Sierra Club applies mainstream approaches in each of its International Committee programs, particularly as seen in efforts to reform the World Bank, it concentrates its energy and resources on grassroots activism. Examples include supporting and mobilizing public activists as well as developing contacts with environmental organizations in other countries. This networking follows a variety of channels. Until the mid-1990s, these simply included face-to-face contacts within a community, official telephone drives, environmental association meetings and conferences, neighborhood association or city council meetings, and clean-up or research campaigns that brought together individuals from all walks of life.[13] With the rise of the Internet, a whole new genus of tools emerged. The medium itself is inherently dynamic. Pages are grouped around a homepage with its own distinct address, technically known as a Uniform Resource Locator (URL), but these pages are constantly changing—sometimes with

Figure 3.2. Sierra Club postcard featuring child with gas mask. SOURCE: Photo: Becky Bornhorst. Michael M. Gunter, Jr., thanks the Colby Memorial Library, Sierra Club for use of its historical archives.

little or no warning. Information available one day may not be available the next. Material may be modified, moved, or even deleted altogether. Every group studied for this book demonstrated this tendency, with several completely revamping their Web sites during the study period.

NGOs do this, of course, in an attempt to keep their material up to date. Material on the Web can become dated just as quickly as other mediums—but this condition may or may not be known to the reader. A large number of Web pages continue to exist virtually but not in reality. A drawback for NGOs becomes how to make their message heard through the plethora of information that bombards society in the twenty-first century, especially when trying to make one issue stand out above an increasingly crowded environmental NGO field. Still, the strengths of this medium largely compensate for its deficiencies. As a graphical, multi-media environment, the Web offers pictures and sound—at times even together in video format. It is an inherently interactive medium of communication. In many respects, it mimics the fluidity of civil society and the character of the NGOs themselves. From Web pages, to mass electronic mailings, to activist discussion listservs, groups clearly take advantage of the potential that instantaneous information distribution brings.

These tools have evolved in the late 1990s to the point that they are now often used in conjunction with mainstream strategies. One cannot simply classify all Internet usage as participatory. What may begin as grassroots networking through Environmental Defense Dispatch, the Ocean Conservancy's Ocean Action Network, or Defenders Electronic Network (DEN), often translates into a direct lobbying tool. As such, many large national NGOs now combine these two strategies to enhance their advocacy agenda. According to Roger Schlickeisen, president of Defenders, the Internet Revolution has forever changed the landscape in which environmental NGOs conduct their business. As he states, "environmental activism will never again be the same" because the Internet allows environmental NGOs to be more effective than ever.[14] It allows groups to reach many more people than they would otherwise, on the order of ten times more than would be possible by mail or phone, he contends. The Internet also is an asset by virtue of the sheer speed at which communication takes place. Threats can be communicated to membership in a matter of hours instead of days. And the Internet simply makes it easier to become active. It increases the ease of response by allowing supporters to e-mail members of Congress and other decision makers with a few clicks of the computer mouse.

Several groups demonstrate incorporation of these points by utilizing an array of electronic instruments. Increasingly, the most popular tool in mobilizing support on a specific issue is the electronic listserv. The Nature Conservancy offers *Great Places* (formerly *Nature News*), a free monthly update on the lands, waters, plants, and animals the Conservancy is working to save. The Sierra Club listserv is *Sierra Club Currents,* which began in January 1997 as the *SF Moderator.* Also known at one point as the *Sierra Club Action Newsletter,* the current incarnation is sent an average of two times a week. Typically, three or four pressing issues are highlighted in each mailing, which encourages recipients to contact their appropriate representatives in Washington as well as local media outlets. Each message concludes with a listing of telephone numbers for contacting the White House, the United States capitol switchboard, and a Sierra Club legislative hotline. The mailing also encourages contact over the Internet and posts e-mail addresses for the President and Vice President and Web addresses for both the House and Senate.[15]

The Ocean Conservancy offers *Ocean Action Network.* Action alerts on this network have asked members to contact the U.S. Commerce Secretary, U.S. National Marine Fisheries Service, and the CITES Convention itself. For instance, in preparation for the eleventh biennial meeting of CITES in Kenya in April 2000, Ocean Conservancy utilized its *Ocean Action Network* to ask its members to send letters to the CITES Secretariat expressing concern that Japan would push six separate proposals to weaken protection for gray and minke whales as well as hawksbill sea turtles. A general form letter was provided, but members were encouraged to paraphrase and add anything else they wished to say. Ocean Conservancy thus applied a combination of grassroots networking and lobbying in its efforts to protect CITES regulations.

Environmental Defense hosts a series of discussion forums (essentially chat rooms) that address biodiversity, business and the environment, environmental justice, global air and climate, pollution and human health, world oceans, and teacher talk. But probably more effective in terms of its broad reach and the partnership dimension it exploits is *Defense Action Network,* a electronic association of thirty-five different environmental groups, including Environmental Defense and Ocean Conservancy's *Ocean Action Network* noted above. Environmental Defense sends e-mail action alerts through this site, which Environmental Defense itself refers to as "a vital force to influence environmental policy in your home state, in Washington, D.C., and around the globe."[16] Azur Moulaert, project manager for *Environmental Defense Action Network,* believes the inaugural year of this network in 1999 was a resounding success.[17] Designed for rapid response, this grassroots network incorporated more than 130,000 activists and garnered Environmental Defense a nomination for a Webby in the newly created activism category for the May 2000 awards.

Known worldwide as "the Oscars of the Internet," the Webby Awards began in 1997 and expanded in 2000 to include twenty-seven categories and again in 2002 to thirty total divisions. Sites are chosen by the International Academy of Digital Arts and Sciences, with a People's Choice Award winner as well. While it did not win the Webby itself, Environmental Defense did receive one of only five nominations in the activism category (sites facilitating political change, social movement, human rights, public education and reform, or revolution) for the *Action Network* site it maintains. *Action Network* links *Environmental Defense Action Network* with thirty-four other organizations and networks that combine grassroots networking with lobbying policymakers.[18] By 2003, over 750,000 activists had taken action with Environmental Defense alone. Activists are encouraged to phone, fax, and e-mail policymakers on specific environmental issues currently under debate. Environmental Defense includes helpful background information on these "action alerts" issues as well as pre-formatted sample letters. Quotations from satisfied users are displayed on the Web site. As one individiual from Dayton, Ohio, states, "As a full-time student, professional and wife, I value *Action Network* because it helps me to quickly contribute to protecting the environment."

Defenders of Wildlife's electronic network, similarly, carries *DENlines,* a free biweekly e-mail newsletter, which was activated in October 1999 and by June 2000 totaled some 130,000 subscribers.[19] During the first four months of 2000, for instance, over 120,000 DEN activists e-mailed President Clinton asking him to reconsider the decision by the Department of Commerce to weaken the "dolphin-safe" label for tuna.[20] Defenders also encouraged its members to mail letters directly to grocers themselves, including thank-you letters to those that pledged not to sell tuna caught by encircling dolphins and pressure letters to those grocers who had not made this pledge.[21] The big question here, of course, is whether such techniques are effective or merely make activists feel good. Defenders believes the former is true and uses *DENlines* regularly to update its members on public policy issues as well as specific congressional actions. Defenders also reports interviews with experts and officials as they relate to particular action alerts. These are all linked together on the Defenders Environmental Network (DEN) Action Center.[22] Archives of all past alerts as well as your own personal history of action are accessible.

World Wildlife Fund (WWF) provides one final example of an NGO utilizing the Internet in its grassroots networking strategy in conjunction with mainstream lobbying. As of June 2003, approximately 31,000 participants in 160 different states have used their Conservation Action Network at one time or another. It allows members to take action easily and quickly, and the free service generates results. Participants living in the United States receive an average of three notices per month about upcoming policy issues, although sometimes this number increases as events dictate. Participants may also go directly to the WWF Action Center Web site to access these alerts. Some four to five online action options are available, each of which has its own form letter ready for activists to send to the appropriate access point, such as the President, members of Congress, state legislators, and foreign government leaders. WWF encourages activists to add their own thoughts and edit the existing form letter if they wish. WWF also asks its members, in the spirit of grassroots networking, to notify friends and colleagues of the action network generally as well as of specific action alerts that may be of interest to them.

The network has proven effective in a number of respects, and was recognized by the Web Marketing Association for both its "standard of excellence" and having the best Web site of any nonprofit organization on the Internet in 1999.[23] One example of such effectiveness is a financial victory in the Galapagos Islands thanks in part to a February 1998 action alert listing. In just over two days, six hundred faxes and e-mails to President Clinton urged him to support President Alarcon of Ecuador in a pending Galapagos Conservation Law. That law was approved by Ecuador in March 1998, providing sweeping protection for this ecologically diverse ecosystem. Then, in October 1998, the U.S. Congress provided $1.2 million for Galapagos conservation efforts. While causation is admittedly difficult if not impossible to prove in this case, networking activity does mobilize a constituency along the lines that are needed for effectiveness to occur.

Another example of grassroots networking contributing to effectiveness is the Rhino and Tiger Product Labeling Act in the United States. Enacted in October 1998, this law was shepherded through Congress thanks in no small part to eleven thousand activist e-mails urging members to prohibit the sale, import, and export of products labeled to contain substances derived from rhinoceros or tigers.[24] Just a few other examples of effectiveness are increased Congressional funding for rhinoceros, tiger, and elephants in April 2000; increased funding for North Atlantic right whale conservation in November 1999; halting of a Russian plan to conduct the first international trade of beluga whale meat in September 1999; and the early 1999 agreement by the world's principal fishing nations to reduce shark over-fishing.[25]

One final action worth note is WWF's reinforcement of the Mexican decision to cancel construction of a giant salt-producing facility near the San Ignacio Lagoon. As former Mexican president Zedillo stated, "Few places in the world resemble the El Vizcaino Biosphere Reserve. It is a unique place in the world for the species that inhabit it as well as for its natural beauty, which is also a value we must preserve."[26] This action was particularly notable because the thirty-four hundred e-mail messages sent to President Zedillo were not in opposition but in thanks for an action that was already taken. These "victories" are prominently posted on the action center Web site along with current initiatives. Direct individual action is just a point and click away. As seen in figure 3.3, the homepage for the World Wildlife Action Center is both practical and visually attractive, with hyperlinks to each of its current action alerts readily available.[27]

Whether simply linking staff members of different state (and often continent) offices within a single organization or soliciting the public at large, organizations repeat a familiar refrain regarding the Web. Like almost all other groups within the 1990s, environmental NGOs recite the mantra of virtual communication by heart. It opens new doors, empowering groups previously left out of the political process. It equalizes access to information outlets. It increases the efficiency of mobilization efforts. It makes the world a smaller place—and highlights the extent to which we are all dependent on one another. Earthwatch Institute believes that these technological advances not only have made communication within its organization easier; technological advances have also made the world itself smaller. Just looking to the grant application process at Earthwatch illustrates this, according to Earthwatch's public relations director Blue Magruder in a personal interview. In one instance, a Czech vet and pathologist wanted to study anti-parasitic plants that orangutans eat in Indonesia. She found out about Earthwatch through the Internet. Professional colleagues, media exposure, and volunteers who never would have been at her disposal now were. Thanks largely to a technology that has been available to the general public only since the mid-1990s, the output of this vet/pathologist quadrupled.

Apollo spacecraft images of the blue, white, and brown Earth struck a common chord in humanity, touching our collective psyche in the early 1970s.[28] In the 1990s, the Internet, especially the high-resolution graphics of the World Wide Web, made a similar impact, sparking an awareness

Figure 3.3. Homepage of World Wildlife Fund Action Center.

that the political boundaries of the past, while clearly still relevant today, mean less and less in the grand scheme of the global issues that confront our planet. The boundaries (if indeed any boundaries exist) of cyberspace, like the environment itself, are very different from those of our conventional political map. They are more permeable and highly interdependent. Moreover, they graphically illustrate the extent to which globalization now lives and breathes in both the virtual—and real—world. Of course, largely because of forces such as environmental and technological globalization, the traditional political cognitive map is also changing. It is becoming less and less boundary specific.

This again is where the unique nature of NGOs, specifically the possibilities of electronic grassroots networking, offers much potential. Yet a

Figure 3.4. Apollo photograph of Earth from space. This view of Africa and Saudi Arabia was taken from the Apollo 17 spacecraft on December 7, 1972. NASA believes it to be the most requested picture of the Earth. SOURCE: Image courtesy of Earth Sciences and Image Analysis Laboratory, NASA Johnson Space Center.

word of caution must be added. Technological advances in communication create new opportunities for oversight as well, opportunities that may run counter to participatory ideals. For instance, Environmental Defense keeps track of response rates to their action alerts by requesting that activists send their response letters directly to Environmental Defense, where they are forwarded to the appropriate policy-making point. Even when not acting as a middleman in this manner, though, Environmental Defense oversees electronic communication in an almost frighteningly Orwellian form. When sent electronically, each action alert has detailed statistics regarding how many activists responded to a particular request, which ones, how many policymakers received those letters, and the number of letters each specific policymaker received. This highlights a darker side to the Internet, and the struggle within environmental NGOs over how much power ultimately should be transferred to their membership—even when engaging in ostensibly participatory strategies such as grassroots networking.[29] EDF could drop individuals from the listserv or potentially even manipulate actual text of messages sent back to them. The point is they retain ultimate control, not the individual.

These concerns about manipulating grassroots networking in an overly controlled fashion are the technological offspring of similar fears voiced by Ernest Yanarella and Sheldon Wolin. These two authors warn that "periodic pseudo-participatory waves of mobilization to write letters, send telegrams or make phone calls" cannot overcome the trend for interest groups to professionalize and assume many of the same characteristics as those interests they opposed, that is, hierarchical decision-making apparatus, mass-mailings, and passive membership. Yanarella addresses environmental groups whereas Wolin looks at peace organizations targeting nuclear war.[30] In any case, future negotiation of these potentially acrimonious power struggles will be critical to grassroots networking effectiveness.

Community Education

Closely related to grassroots networking, but worthy of classification as a strategy in its own right, is the strategy of community education initiatives. Community education efforts can take the traditional form of pamphlets and voter education, as evidenced in the one million voter guides printed by the Sierra Club during the 1998 elections.[31] They can also assume more imaginative forms, such as the Sierra Student Coalition's Rainforest Education bus. Used to raise awareness about the Great Bear Rainforest on the central coast of British Columbia, this bus traveled throughout the eastern United States during the spring of 1999.[32] Still another example is Sierra Club's 1999 "Earthrise" global warming postcard campaign, which distributed fifteen thousand postcards highlighting the dangers of global warming in a matter of weeks. Addressing the impact of global warming on species loss, this postcard featured the famous "Earthrise" photo from the Apollo 8 moon mission. Global warming is directly related to biodiversity issues, because it reduces the habitat available to species and threatens some of the most susceptible populations in the world, often through a mechanism termed the mountain effect. The mountain effect refers to the splicing of habitat by either basic human encroachment or the more complicated phenomenon of global warming. The result is that sub-populations of species such as the giant panda are trapped in isolated areas like the tops of mountains without access to the rest of their habitat range and the rest of the their population members. Genetic diversity within the species, and thus species diversity itself, is then threatened.

Members were asked to address this postcard to their respective U.S. Senators. On the back of the postcard, requests were made to raise miles-per-gallon standards, curb power-plant pollution, and invest in alternative energy sources such as wind and solar power. Activists distributed these postcards in different ways. Jim Wallace, a science teacher in New Jersey, used the postcards in his class to help teach a unit on global warming and to involve his students in the democratic process. Others such as Florida activist Jennifer Caldwell included the postcards in her group's monthly newsletter. The response by members was much better than project organizers expected, and the Global Warming Campaign was happily forced to print a new batch in 2000.

Earthwatch Institute provides perhaps the best institutional example of education initiatives as a participatory approach. Founded in 1971, Earthwatch now claims fifty thousand supporters globally and maintains offices outside the United States in Oxford, England, Melbourne, Australia, and Tokyo, Japan. Each year, the NGO sends roughly 3,500 members to work with more than 250 research scientists on some 140 projects in a total of 50 different countries across the globe. These projects, which address biological diversity, cultural diversity, and general world health and historical origins, promote sustainable conservation of our natural resources and cultural heritage by creating partnerships between scientists, educators, and the public at large. Earthwatch has completed projects that directly supported the discovery of two thousand new species, establishment of fifteen national parks, and the founding of nine museums. As stated on the opening pages of their annual expedition guide, Earthwatch places a fundamental faith in the average citizen. The organization believes that "sustainability can be achieved only if ordinary people . . . [are] actively involved in solving the world's problems." This occurs, Earthwatch believes, "by bringing the people to science."[33]

Earthwatch combines scientific research with community education. Since their founding, Earthwatch has operated according to this two-tiered approach, with the third dimension of conservation added in recent years. In their literature, Earthwatch lists conservation as a separate objective, on par with research and education, and in terms of end result this is true. Of course, this book asserts, from a strategy standpoint, that both education and research are aimed at conservation by definition and to list it separately would be redundant. More importantly, Earthwatch has become more active over time in seeking partnerships with other organizations so that conservation on the ground takes place. With this in mind, in 1999, Earthwatch began entertaining the possibility of making experiential education the central part of their role. Interestingly, the over-arching educational goal of Earthwatch is also twofold in that it seeks to help raise both the level of public awareness generally and the level of personal responsibility specifically. Their ultimate objective is a sustainable future. Recognizing this strategic niche, Earthwatch foresees its role as "waking people up to the interconnectedness" of the world. Much as E. O. Wilson and Stephen Kellert have argued, Earthwatch believes that "biophilia" exists within people, that ecological responsibility will grow once this love of nature is woken from its deep societal slumber. The organization wants to change the way people read newspapers, vote, and recycle. It wants people to volunteer. Quite simply, it seeks to "change the world, one person at a time." [34]

Earthwatch induces this change during expeditions lasting between one to three weeks, during which members serve as volunteers assisting world-renowned scientists in their data collection. Typically projects utilize between eighteen and fifty volunteers with a ratio of approximately one scientist to every five volunteers. With five to twelve volunteers on each team, the group spends ten to twenty days in the field. Each participant pays a share of the costs of the expedition, at an average cost of $700 to $4,000 per individual, excluding airfare. [35] Each participant also does a share of the work for the expedition. Volunteers perform a myriad of tasks, including identification of what species live in an area, what nests where and when, and what trees as well as other plant matter are needed to sustain life. This data is then used to establish broad baseline studies and make evaluations of the health of various species populations. It is also useful data for advocacy organizations seeking to give input on specific policy issues, such as whether just a rogue coyote kills sheep or if all coyotes kill sheep. Earthwatch thus provides data to answer policy questions.

Earthwatch research falls into two broad categories. There is basic research intended for future applied researchers to utilize, and there is research that is specifically geared toward conservation efforts. In both cases, research sponsored by Earthwatch tends to be a bit more daring and unconventional than government-sponsored work. With the freedom to engage in more nontraditional research, scientists who have received Earthwatch grants consistently praise the more creative and experimental environment promoted by Earthwatch, one that allows scientists to be more aggressive in pursuing their work.[36] Earthwatch also funds research for other nonprofit organizations that sometimes need additional funds to complete baseline data, including Massachusetts Audubon, The Nature Conservancy, and Conservation International. This typically includes labor-intensive research, as in work with The Nature Conservancy in the old-growth forests of California here in the United States or assistance provided to Conservation International scientists on their survey of the Togian islands of Indonesia. All told, Earthwatch is second only to National Geographic Society as a supporter of field research, with individual grants averaging about $25,000.

Still another example is the Bahamian Reef Survey conducted by marine biologists Thomas McGrath and Garriet Smith. For over eleven years, these scientists have studied coral reef decay and bleaching—with the help of Earthwatch volunteers such as the one shown in figure 3.5. Snorkeling over San Salvador Island, one of the Bahamas outermost islands, volunteers not only get an up-close and personal view of coral reef degradation, they help conduct research that will provide data to prevent future bleaching and disease. Bleaching occurs when the algae that normally live symbiotically with the coral die due to environmental stress. That leaves a colorless, and more importantly, nutrition-less coral reef. Volunteers are in the water four to five hours a day systematically studying the health of the reefs, then spend the evening transcribing data, cleaning equipment, viewing films, or hearing presentations about the region.

From swimming with sharks in the Caribbean to counting kangaroo droppings in Australia, Earthwatch volunteers provide invaluable assistance. Utilizing volunteer research assistants is not without drawbacks, however. Chief among these is the virtually omnipresent concern regarding the reliability of data gathered by volunteers.[37] Initial reaction within the scientific community to this practice was very guarded, but much progress has been made since the 1970s in addressing methodological concerns.[38] An emerging trend in scientific research more generally is wider incorporation of volunteer assistance. This in turn supports efforts at Earthwatch, particularly because the organization strongly believes that controversy implies the need for more information. Earthwatch does not engage in direct politics. It does not lobby or advocate specific policies. The organization has no offices in Washington and is not a major force to provide scientific results to government or to the general public. Instead, Earthwatch is an experiential organization, emphasizing the participatory strategy of education, and to an extent grassroots networking, in combi-

Figure 3.5. Earthwatch volunteer snorkeling for data. SOURCE: © Elizabeth Brill/ Earthwatch.

nation with scientific research. People experience personally the ecological relationships, the need for biodiversity protection.

In some respects, the strategic niche that Earthwatch carves out by combining the mainstream strategy of scientific research with the participatory endeavor of community education reflects a broader philosophy that the entire environmental community would be wise to adopt. Earthwatch's approach is a function of the increasing recognition that public awareness is a key factor in conserving biological ecosystems. Several of its Web pages reflect this, as teacher grants and specific lesson plan ideas dot the landscape. The Earthwatch Global Classroom, for instance, offers not only expedition profiles but also interviews with the very scientists that lead these expeditions. It allows teachers to assimilate Earthwatch material directly into their curriculum, particularly as periodic "virtual field-trips" combine with other sites such as Scholastic Network and Discovery On-line to facilitate more active learning.[39] This type of outreach as well as the periodic, Earthwatch-sponsored scientific lectures and community educational events together create a genuinely more enlightened global citizenry.[40]

But the impact of Earthwatch can also be seen in indirect respects. Staff members may serve as intermediaries for various government officials as they recommend scientific contacts to members of Congress who seek advice on proposed legislation. Earthwatch also feeds its results into the work of other organizations. At times, material from its scientists research ends up in the Sierra Club manual or other publications, for instance. At other times, research results are used by others in actual mainstream lobbying strategies, providing the scientific grounding for a particular lobbying initiative. A good example of this is the eight years of research Earthwatch funded on barrier islands, under a grant investigating whether beach buggies destroy sand dunes. Detailed data gathering determined that, as long as beach buggies stayed where they were supposed to stay, no appreciable damage was felt. Instead, researchers found the one regulation that would best enhance dune protection efforts would be requiring people to go barefoot. The logic behind this is simple: People wearing shoes trample grass, even if unintentionally, whereas humans with bare feet try to avoid scrapes and scratches. They walk carefully. This alone would make the biggest difference in protection efforts. The National Park Service decided against such an "intrusive" regulation, however.

This story is particularly illustrative as it relays the conversion of a previously skeptic observer of Earthwatch activities. A senior Sierra Club official once told Blue Magruder, public relations director at Earthwatch, that, to be really effective, environmental groups had to be in Washington, D.C. He argued that simply gathering research would not have any major impact. Years later, though, that same individual approached Magruder and said he was wrong. The Sierra Club was actively pursuing barrier island legislation that would restrict FEMA assistance if someone built her house in an undeveloped area and then had it destroyed in a natural disaster. This individual directly attributed legislative success on the proposal the Sierra Club supported to congressional testimony from Steve Leatherman of the National Park Service, the same Steve Leatherman who Earthwatch funded for eight years of barrier island and dune research. The lesson here is that strategies may even support each other when no such partnership was intended—and this symbiotic effect may be felt years after original implementation. This is the overarching goal to which Earthwatch strives with its educational (and research) emphasis.

Finally, educational initiatives targeted at the environmental NGO community itself deserve mention. Biodiversity Action Network (BIONET) devoted substantial energy to educational efforts in this arena. While some of this involves networking, such as organization of a roundtable bringing together people working on climate with those working on biodiversity, the central goal of BIONET was undoubtedly raising awareness. One of the core functions of the organization, as was discussed in chapter 2, was and still is its information clearinghouse. Another example was the well-received, user-friendly guide that BIONET produced for the Jakarta Mandate on Marine and Coastal Biodiversity in November 1995. The network of NGOs that was BIONET produced a variety of other documents as well, including a 70-page NGO study guide to the Global Environment Facility (GEF) co-produced with IUCN and Climate Network Europe.[41] Documents themselves are not the only manner in which BIONET contributed to educational efforts. BIONET often served as an informal leader in various diplomatic settings, the first Intersessional Meeting on Operations of the Convention (ISOC-1) on Biological Diversity in Montreal, Canada, for one.[42] During the initial two days of this three-day meeting, BIONET coordinator Hans Verolme assumed a lead role in brief, morning strategy sessions organized by attending NGOs. Here group representatives would discuss the content of work in the plenary sessions and various working groups as well as specific individual states that might be most likely to entertain NGO positions. The NGOs then returned to observe members states in action at the plenary sessions themselves.[43]

Summary and Findings

These two participatory strategies then combine clearly to enhance transnational biodiversity protection efforts by environmental NGOs. Grassroots networking brings together individuals from throughout a state, and, at times, throughout the globe. This is a particularly powerful tool because it fosters communication between groups that otherwise would

Table 9. Strengths and Weaknesses of Participatory Strategies

	Strengths	Weaknesses
Grassroots networking	Connects previously disjointed (and sometimes disparate) groups to form a whole that is more politically viable than the sum of its individual parts; fosters future efforts by laying foundation for developing solid constituency of either support or opposition.	NGOs lose direct control over an initiative and may then lose focus due to the sheer number of actors (and thus interests) involved.
Community education	Enhances the political viability of individual activists and, as with grassroots networking, creates potential for partnerships among what were previously disparate groups	Can be expensive; also, groups often try to classify what is simply a fundraising activity as an educational initiative so the line between facts and political advertising has been blurred by NGOs themselves.

never contact one another. Bringing together previously disparate individuals and the organizations they represent enhances not only the political power of each individual, it also increases the probability that their programs will be effective. Similarly, community education initiatives enhance the political viability of the civil society that environmental NGOs utilizing participatory strategies seek to mobilize.

Together with grassroots networking, then, community education initiatives allow NGOs to define an issue; mobilize membership, staff, or the general public; access decision-making points; and implement specific projects. These strengths must be couched within the context of several weaknesses, though. Participatory strategies, particularly educational initiatives, can be expensive. Also, while both grassroots networking and community education enhance the potential for long-term results, they also take much longer to develop than most mainstream strategies. Grassroots networking also runs the risk of creating a confusing array of actors and interests when it opens up the political process to direct participation within a membership. Environmental NGOs tread a thin line as they seek to maximize the benefits of participatory democracy, but minimize the costs that may arise in the form of a diluted focus. The strengths and weaknesses of both grassroots networking and community education are summarized in Table 9.

FUNDAMENTAL LINKAGES

Domestic-International Linkages

Whether utilized as separate strategies or in tandem, grassroots networking and community education initiatives provide a critical function by drawing connections between the people of a community and the actual species protection policies themselves. They allow issues to develop resonance within a society—and give NGOs added political power in their efforts to establish the three fundamental linkages. While many tools are needed in constructing the ark of the twenty-first century, participatory strategies may be the most important. Nowhere is this more evident than in the realm of domestic links to international biodiversity protection and vice versa. Forging this linkage is not an easy task. To date, many environmental NGOs struggle with even just the rhetorical components as they debate renowned futurist and evolutionary economist Hazel Henderson's advice to think globally but act locally.[44] Perhaps most importantly, environmental NGOs too often skip the stage of building a domestic constituency for solving a global problem. As Dan Seligman, international trade and the environment research specialist at the Sierra Club, cautions, this is a critical mistake.

Looking to the Biodiversity Convention and its inability to take hold within the United States provides valuable insight on this failure. Precisely this lack of a domestic-international linkage dictated events as the U.S. environmental community lost focus despite an initial consensus regarding the need for domestic-international connections. After paying rhetorical credence to linking the two, groups rushed into international negotiations. They rushed to line up international partners for assistance in their work. They rushed to refine their diplomatic approaches. But nobody ever made a compelling case for the domestic side of the equation.

According to Dan Seligman, the environmental community, including the Sierra Club, simply "dropped the ball" on this one.[45] The community made failure essentially inevitable when it neglected to integrate the three key ingredients in any fight: money, people, and the media. None of these three components were developed or deployed in the interests of the Convention on Biological Diversity. Nobody ever formulated a strategy. To a certain extent, this may be explained by the monumental changes that preoccupied most policy-making circles as the decade of the 1990s began. Without the larger Cold War context, the United States lacked a dominant guiding rationale for its foreign policy. The environmental community, much like many other interest groups within the United States and abroad, was not prepared for this dissolution. Certainly, environmental NGOs were not trained to recognize the larger arena prior to this event. Environmental NGOs thus were woefully unprepared to exploit this foreign policy vacuum, a lack of readiness painfully evident in the negotiations to the Convention on Biological Diversity (CBD) at the Earth Summit in Rio de Janeiro, Brazil, in June 1992.

Without a unifying force to fill the void created by the end of the Cold War, without a guiding U.S. foreign policy rationale, former President Bill Clinton spent the eight years of his presidency promoting a global vision as his centerpiece, albeit with a fair number of domestic scandal sidetracks. While his administration fought foreign policy fires when they broke out in a security sense, the dominant agenda was to foster alliances with business leadership interested in exploiting these international angles. The environmental side effects to this approach were often glossed over or left out entirely. From trade rules themselves to the impact that foreign investment has in terms of opening up resources to extraction and unsustainable application, environmental concerns took a back seat in many of the development initiatives of that decade.

The North American Free Trade Agreement (NAFTA) is a prime example, and NAFTA quickly became a polarizing issue in the environmental community. In very simple terms, people did not know what the rules would mean in 1993. They were mysterious, poorly written regulations, so legitimate disagreement naturally arose. As the close of 1999 approached, most NGO representatives felt that progress had been made in this arena and that an essentially unified, and dignified, environmental front would help shape international trade more favorably. The outbreak of violence at the World Trade Organization (WTO) summit meeting in Seattle in November and December 1999 and subsequent postponement of the WTO's meetings combine with more recent protests to highlight the extent that vociferous labor and environmental disagreements remain, albeit often drowned out by a minority of anarchists and their attention-getting tactics.

The U.S. environmental community reacted complacently in the context of the end of the Cold War, thinking that previous victories with the Clean Air Act and Clean Water Act in the United States had already established them as a force to be reckoned with on the policy-making battlefield.[46] But environmental activists misunderstood both the scale and the very nature of the challenge they faced. During the 1990s, the Sierra Club made a concerted attempt to reverse this process, to correct for the gross deficiencies in the standard approach. Mobilization as such became the primary objective of the Sierra Club, both domestically and internationally.

This approach directly applies what Seligman believes to be the only well of countervailing power available: a combination of money, media, and, perhaps most importantly, people.

Now with President George W. Bush and a more conservative administration as well as a Republican-led Congress, this perspective is all the more important. As Tom Turner, senior editor at Earthjustice, laments, "Right now we have no friends in high places. There are some good people in the House, but they are quite outnumbered."[47] Perhaps the shift to the right inside the Beltway, though, will have a positive spin for the environment. Membership skyrocketed across the board in the 1980s, for example—precisely at the same time that the Reagan Revolution took hold. And the globalized world of today means that many options to influence U.S. policy may exist from outside our borders as well. By the time of the United Nations' World Summit on Sustainable Development in Johannesburg, South Africa, in August and September 2002, for example, world opinion certainly had crystallized against the United States, and the Bush Administration in particular, at least regarding environmental issues. As Steve Mills, director of international programs at the Sierra Club, notes, much of the world sees the issue of environmental protection differently than the United States government. And perhaps foreshadowing the French-German-Russian split with the United States over its decision to invade Saddam Hussein and Iraq in March 2003, foreign delegations showed no hesitancy to speak their mind and offend the American superpower.[48] Mills asserted, "The degree to which our allies were willing to rebuff the Bush Administration on these issues was stunning." Mills also believes that the Bush Administration is oversimplifying international environmental politics. He thinks that foreign policy options are falsely being framed as those that pit economic development against environmental protection, that you cannot have one without limiting the other. As he explains, "The story in Johannesburg was business versus business. The Bush Administration is listening to old polluters, the old technology of oil and coal."[49]

By focusing on people themselves, the Sierra Club positions grassroots networking above all other strategic options. Applying a style that had almost left American culture altogether, the Sierra Club calls to mind the age-old complements of the famous French student of American democracy, Alexis de Tocqueville, and reconnects political efforts to the traditional interest group approach of American politics. Nothing illustrates this better than the "activist toolkit" the Sierra Club offers on its Web site. Here one may find a series of tips for effective activism, including detailed advice on writing a persuasive letter to elected officials, distinctions to con-

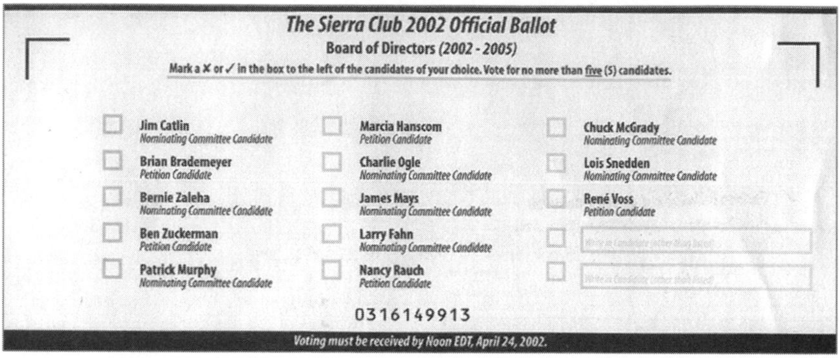

The Sierra Club 2002 Official Ballot

Board of Directors (2002 - 2005)

Mark a ✗ or ✓ in the box to the left of the candidates of your choice. Vote for no more than five (5) candidates.

☐ Jim Catlin *Nominating Committee Candidate*	☐ Marcia Hanscom *Petition Candidate*	☐ Chuck McGrady *Nominating Committee Candidate*
☐ Brian Brademeyer *Petition Candidate*	☐ Charlie Ogle *Nominating Committee Candidate*	☐ Lois Snedden *Nominating Committee Candidate*
☐ Bernie Zaleha *Nominating Committee Candidate*	☐ James Mays *Nominating Committee Candidate*	☐ René Voss *Petition Candidate*
☐ Ben Zuckerman *Petition Candidate*	☐ Larry Fahn *Nominating Committee Candidate*	☐
☐ Patrick Murphy *Nominating Committee Candidate*	☐ Nancy Rauch *Petition Candidate*	☐

0316149913

Voting must be received by Noon EDT, April 24, 2002.

Figure 3.6. Official ballot for Sierra Club board of directors. Every member is entitled to vote in these elections. This ballot lists thirteen candidates with two spots available for write-in candidates. Members voted for up to five individuals. Michael M. Gunter, Jr., thanks the Colby Memorial Library, Sierra Club for use of its historical archives.

sider when that letter is a letter to the editor, and point-by-point recommendations to prepare, present, and follow up on a meeting with elected officials. There is even a section on "how to raise hell at public hearings and community meetings."[50]

Participating in the political process in a thorough and sophisticated manner, the Sierra Club personifies a textbook case of middle-level civil society at work. It gives people confidence. It gives people the tools for good citizenship. Whether it is making signs to get attention or simply turning people out to vote, both development and maintenance of citizenship are central to the Sierra Club's mission. As Seligman succinctly states, "We believe in citizenship at the Sierra Club."[51] Membership reflects this, as savvy, pragmatic individuals commit themselves to the political process. Unlike many greens, members within the Sierra Club certainly are not afraid of using power. They embrace political power and pride themselves on being knowledgeable about the issues. Nothing demonstrates this better than the annual elections to the Sierra Club board of directors. In this respect, the Sierra Club is the most democratic of the large national environmental NGOs.

All this relates to our domestic-international linkage because it addresses the need first to establish U.S. domestic awareness on the issue of transnational biodiversity loss if the average U.S. citizen is to care about relevant international treaties such as the CBD or CITES. This recognition makes a profound difference in the policy programs that an organization proposes. As Seligman explains, "If the first question when you come to work in the morning is how do I get people involved, you do very different things from if you ask yourself how do I save biodiversity." Any group can take advantage of this approach. Of course, it helps to have the tools that the Sierra Club has at its disposal, namely integrated communication vehicles and structured membership as well as experienced and outgoing people. The Sierra Club boasts an organizing culture that attracts people who are not timid, who display a growing commitment to environmental issues they hold dear. While the Sierra Club has both experience through its age as well as the weight of size to throw around, an NGO does not have to be older or develop a large staff to implement grassroots networking effectively. Even small staffs of just two to three people can employ grassroots networking with great success. Public Citizen Watch, for one, has built a tremendous grassroots network that serves as a foundation for future conservation efforts.

Drawbacks to the decentralized approach that the Sierra Club employs certainly exist as well. Often there is no clear decision maker. Issues can go around and around, taking longer to frame than if a more centralized process existed. Significant political pitfalls may also arise within this participatory process, as seen in the 1998 immigration vote at the Sierra Club. In 1998, a membership ballot initiative, proposed by a conservative, America First element within the organization, was soundly defeated 60 to 40 percent. Racist overtones permeated the discussion and other environmental groups were rightfully offended. Some fallout continues today, but Seligman believes that offended groups are beginning to realize that the majority of the Sierra Club, including a leadership that spent large sums of funds campaigning against the initiative, does not favor this exclusive ideology. Despite the short-term costs, then, this immigration vote example also offers encouragement that the difficulties presented by decentralized decision making can be overcome without compromising participatory values. Seligman believes that the participatory headaches and handicaps are worth the benefits they provide in the long run. The real challenge for the future will be to increase the scale of what the Sierra Club is doing. Borrowing from marketing lingo, a demand-driven model helps formulate mission objectives in this scenario. It makes possible development of a viable domestic constituency for international issues.

WWF also employs grassroots networking to support the domestic-international linkage, and, like the Sierra Club and The Nature Conservancy, WWF sponsors various tourist expeditions that attempt to instill networking as well as fundamental educational aspects.[52] Trips such as WWF's Tanzania Wildlife Safari are led by respected naturalists and seek to inform participants firsthand about WWF's conservation priorities. That is, what might first superficially appear to be little more than glossy tourist brochures are actually intended to create the very constituencies that the Sierra Club's Seligman pointed out as lacking.[53] WWF wants to show people in states on the other side of the globe why they should care about what happens in Tanzania and other states with valuable biodiversity reserves. WWF's Travel Program now includes trips to the Amazon, Alaska, Belize, the Northern Rockies, Iceland, and Tambopata, among others.[54]

Of course, these tourist initiatives typically attract only individuals from rich, Northern hemisphere states—but domestic constituencies exist in southern states as well. Radio WWF is one example of a project that recognizes this and attempts to bridge the gaps that exist there. Heard in nineteen Latin American countries, Radio WWF takes advantage of the fact that radio is often the primary medium of communications for rural areas in Latin America. Some three hundred stations daily broadcast various environmental scripts designed to help citizens improve their quality of life along with their environmental surroundings. Surveys show that most stations have adapted the scripts to their specific programming needs, and WWF hopes eventually to expand the total number of stations on the network to five hundred. The project, which is run out of the WWF office in Colombia, is a direct application of the educational strategy.[55]

Another example of grassroots networking that attempts to draw connections between local and global constituencies is the aforementioned International Coastal Cleanup campaign conducted by the Ocean Conservancy. Recruits volunteer for this annual event held on seashores throughout the world. According to Bob Irvin, former vice president of marine wildlife conservation and general counsel at Ocean Conservancy, these volunteers are enthusiastic and a highly effective tool in grassroots networking. For some of these volunteers, it is the one thing they do for the environment every year. For others, it is the first step into a lifetime commitment to activism. Grassroots networking builds the framework for future policy and protection successes. It enhances the effectiveness of the domestic-international linkage.

Earthwatch Institute provides another such example, as the organization makes every effort to work closely with those from other countries. Many researchers on projects are native to the area of study, which obviously makes them more effective in getting their study information used in practical applications. In fact, numerous field research fellowships are awarded to native conservation workers in Africa and Asia precisely because this helps increase their effectiveness. If the original researcher is not from the host country, Earthwatch requests local assistance. Earthwatch

Figure 3.7. Working for local experts in Inner Mongolia. SOURCE: Ed Wilson/ Earthwatch Institute.

also makes it standard policy to budget funds for staff to be hired within the local area of each project. Figure 3.7 shows an American volunteer in Mongolia, China, recording topographical and hydrological data under the direction of a Chinese field assistant. Directed by Professor Wei-Zu Gu of Nanjing Institute of Hydrology and Water Resources, the project seeks to identify the source of desert groundwater in the Gobi and the rate at which that water table is dropping, so that more sustainable management of that resource can occur. Desertification impacts biodiversity and is a problem worldwide, but particularly in China where an "area the size of Poland is prone to desertification."[56]

Another specific example of this approach is the work on large predators in Nepal being conducted by the University of Minnesota's David Smith. Smith sought to develop strategies to protect tigers that were too big to keep in the relatively small confines of the national park territory already set aside. Earthwatch volunteers, many of whom were American citizens, worked closely with Smith and Nepalese researchers in examining alternatives to simply enlarging existing park territory. Options developed included corridors, roads, and underpasses—all within a detailed analysis of the closest appropriate distance between these forest linkages and human development. This research was not just for international benefit. Local interests were paramount, and nobody was allowed to see the reports before the head of Fish and Wildlife for Nepal. This process recognized power tensions in the project and the potential for imperialistic overtones if local considerations were neglected. The end result is that a valuable trust developed between Earthwatch researchers and their Nepalese colleagues. Implementation became a local idea instead of being dictated by an outside authority.[57]

Much is at stake here. Steeped in the theoretical tradition of both interdependence and globalization, the domestic-international linkage is perhaps the most complicated of all linkages that NGOs seek to bridge. Fundamental realist concerns clash with liberal requirements for cooperation as a truly global commons finds itself threatened without proper acknowledgement of that threat—and without proper acknowledgement of the responsible parties. Fortunately NGOs offer much promise in mitigating these obstacles as they operate both above and below the state. NGOs exhibit the most potential to function, both rhetorically and physically, as the instruments that make these linkages possible.

The participatory strategies of grassroots networking and community education add to the five mainstream strategies discussed in chapter 2 along these lines. Participatory strategies enhance the effectiveness of NGOs in meeting this linkage, namely by improving the ability of NGOs to define the issue in a domestic-international context and take advantage of the access points that mainstream strategies help create. Still, even with participatory strategies, NGOs are woefully short of achieving an effective linkage between domestic and international considerations. As with the mainstream strategies, fundamental deficiencies in mobilization and implementation exist. Mobilization is the most critical missing link here, for it serves as a prerequisite to the ultimate goal of implementation. Environmental NGOs must concentrate their efforts on piecing together a viable domestic-international constituency. They must mobilize staff, membership, and the general public before they can meet this linkage requirement and be truly effective. One step toward making this mobilization more likely is to address our second linkage, which connects ecological and economic considerations.

Ecological-Economic Linkages

In many ways, linking domestic and international concerns is dependent first on linking ecological and economic concerns. From a realist perspective, people in one area of the world only care about what happens in another area of the world when it directly affects them, and framing bio-diversity protection within the ecological-economic linkage exploits pre-cisely this characteristic. As we shall see, though, this approach also en-counters some fundamental definitional obstacles because biodiversity holds different meanings within different constituencies. Development lies at the heart of this debate. Making economic progress without compro-mising our ecological future is a highly contentious issue, particularly as the very act of measuring development itself is often called into question. Here again, NGOs struggle with the first indicator of effectiveness, issue definition. Complicating matters even further, economic worth must be established both at home and abroad, or perhaps more accurately stated as locally and globally. Doing this locally is tough enough. Globally it is even more problematic, since accurately defining worth often requires es-tablishing constituencies in areas far from where biological diversity phys-ically lies.

As with the domestic-international linkage itself, NGOs attempting to bridge ecological-economic concerns must first deal with erroneous, pre-conceived notions about what is the ultimate self-interest of a state and particular community. Some progress has been made at the local level. Thanks largely to participatory strategies, the mass public now recognizes biodiversity as a key tool within the pharmaceutical industry. The medical advances of today are largely derivatives of the diversity of organisms on Earth, particularly the tropical plants within developing states. As dis-cussed in earlier chapters, Madagascar's rosy periwinkle as a cure for child-hood leukemia is one such example. Similarly, agricultural interests have an economic stake in biodiversity, as many crops of today are genetically en-hanced through wild species. Genetically diverse wild species provide in-surance against threats of disease and pests that typically attack the more uniform, mass-produced crops. Continued support for these wild species, thus, directly targets the ecological-economic linkage.

World Wildlife Fund's (WWF) Zimbabwe program addressing safari hunting provides another example of how NGOs address this linkage. WWF puts money directly in the hands of local people to develop conservation agendas instead of channeling it through central governmental bodies. Local villagers then have a direct economic stake in conservation instead of supporting illegal poaching operations. Quite simply, habitat that species need to survive gets saved when it has economic value to those that live in or around that land. Conservation International (CI) provides still another example of this basic fact in their support of the cocoa farmers of Ghana. CI believes that projects like this must be participatory. That is, key stakehold-

Figure 3.8. Farmers drying cocoa beans in Ghana. The 30,000-member Kuapa Kokoo cocoa grower's cooperative works with the government of Ghana and CI to connect conservation with farming. The project began in the shadow of Kakum National Park, part of the Guinean Forests of West Africa biodiversity hotspot. SOURCE: Conservation International Annual Report, 19. © John Buchanan/ Conservation International

ers, including government and community members, must be involved. Almost one-third of Ghana's national revenue and one-half of its employment depend directly upon the crop of cocoa. By educating villagers about available organic techniques and methods to increase productivity, CI facilitates plant husbandry improvements. It draws direct economic connections to the ecologically diverse habitat of an area. Working with local people by researching new agricultural methods to replace agrochemicals and encouraging the inter-planting of cocoa with other food crops (including planting cocoa under the shade of other cash trees), CI assists communities in setting aside more habitat to be saved.[58] It links ecological and economic concerns.

CI is also notable for its work in eco-tourism. Eco-tourism involves not just responsible travel to natural areas, it also requires directly involving local people so that their welfare can be improved. This, in turn, provides local people with an incentive to support conservation in an area.[59] As noted in the previous discussion of WWF, eco-tourism contributes to other linkages besides the ecological-economic one, notably the domestic-international linkage. Eco-tourism is perhaps most useful in an immediate sense, though, for its contributions to building a local constituency that has a vested economic interest in protecting surrounding ecology.

Beginning in 1989, CI recognized the need to create economic opportunities for local residents, not just for the large national or even international tourist agencies that usually conduct safari expeditions. By acting as a liaison between the tourist industry and local communities, a number of CI ventures have made significant inroads toward solidifying the ecological-economic link. CI works with local tour operators in at least seven countries where hotspots are located, helping to build awareness about the environment as well as improve the economic stature of their specific communities. One such example is the combination of the Scarlet Macaw Trail and the community-owned Spanish language school known as Eco-Escuela de Espanol, both in Guatemala's Peten region. Once the center of Mayan civilization, this area also boasts a famous breeding ground for the endangered scarlet macaw. The CI-initiated project offers students a non-tourist environment in which to learn about the language, culture, and ecology of the area. The financial benefits of the program go directly to the local people. And it also relieves pressure on deforestation. The initiative clearly performs a variety of functions, then, but it is most noteworthy for the extent to which it expands economic opportunities for those who actually live in the Maya Biosphere Reserve area.

CI also hopes to utilize eco-tourism to draw the economic-ecological link on the northeastern shoulder of South America, in Guyana. A biologically rich country of 83,000 square miles, Guyana only has one national park. And this park, Kaieteur National Park, is essentially unprotected. Nestled along the Potaro River, Kaieteur was established in 1929 by the British Commonwealth and originally included 44 square miles. In the early 1970s, this area was slashed to just 7 ½ square miles before 1993 legislation targeted expansion of the park to 224 square miles, at least in theory. Management remains wanting and the government must better build public interest in supporting eco-tourism. That is certainly feasible given that the crown jewel of the region, Kaieteur Falls, is nearly five times the height of Niagara Falls.

CI is helping do just this with its documentary, "A Dream for Guyana's Heritage." The film expressly targets the people of Guyana, attempting to create momentum to end the country's reign as the only Western Hemisphere state without its own protected area system. It has been shown in numerous settings, including the prestigious Jackson Hole Wildlife Film Festival, where it was nominated as a finalist in 1999.[60] A series of six public service announcements for Guyanese television also was a finalist at the Jackson Hole festival, and, together, they are an imaginative tactic in CI's arsenal of community education strategies. As Haroldo Castro, vice president of international communications at CI, states, "Everything we do in my department is to raise awareness. Guyana is still the only country in Latin America without a protected area system. We launched this documentary and the PSAs to provide support for the government and to push government to create one."[61] The in-house editing facilities at CI allow them to pursue precisely this strategy in a cost-effective manner. CI has produced some five hundred hours of professional-quality footage in Castro's twelve-year tenure at CI and holds an impressive video library to prove it, winning more than twenty awards at international film festivals.

Environmental Defense also incorporates the ecological-economic linkage in the majority of its initiatives. While their activity in lobbying the World Bank was documented in chapter 2, participatory strategies such as grassroots networking also apply in this case, as seen in a June 2000 action alert sent by Korinna Horta, senior economist in the International Program at Environmental Defense.[62] Horta asked activists to send faxes to then U.S. Treasury Secretary Lawrence Summers. She sought to influence the vote by the World Bank board of executive directors regarding a $300 million loan for an 670-mile oil and pipeline project beginning in the Doba oil fields of southwestern Chad and ending in the port city of Kribi, Cameroon. Environmental Defense believes, that given the history of instability, civil war, and official corruption in these two African states, the economic benefits promised to local communities from this project will never be realized. Of course, the project now underway also will inflict sizable damage on the fragile ecosystems of both states, particularly the tropical rainforests in Cameroon where the southern portion of the pipeline would pass. The project is the single largest investment in Africa today, with an estimated total cost of $3.7 billion. Horta believes that this is but the latest in a long string of World Bank–sponsored projects that provide support for authoritarian regimes and further marginalize the poor as well as threaten biological diversity.[63]

Environmental Defense also employs grassroots networking through non-Internet means, namely regular postal mailings to its members. One such example is the Clean Car Campaign, which calls for environmental improvements in automobile construction and usage, from their production and operation to their retirement. Environmental Defense seeks to reduce use of toxic paint and toxic heavy metals, increase use of recycled materials, and improve fuel efficiency. This last component, in particular, relates to biodiversity protection, because greenhouse gas emissions contribute to global warming, which in turn increases the likelihood of flooding, droughts, destruction of wetlands, and spread of infectious diseases. All of these threaten not only human health; they threaten all species. Environmental Defense applies its membership network here with a twist on the mainstream approach of lobbying by asking members to send notices to General Motors Corporation, Ford Motor Company, and DaimlerChrysler. These notices state that the activist will choose his or her greenest option the next time he or she is in the market for a vehicle and points out that Honda and Toyota already offer such choices. Environmental Defense, thus, exploits the ecological-economic linkage.

Purely educational initiatives also utilize this linkage. World Resources Institute, for instance, was an important actor in shaping the text *Global Biodiversity Assessment*. Funded by the Global Environment Facility (GEF) and the United Nations Environment Program (UNEP), the *Assessment* represents the work of a collection of thirteen teams of experts with some three hundred authors from fifty different states. It was formally presented in November 1995 at the second COP to the CBD in Jakarta, Indonesia. Totaling 1,140 pages, the report focuses upon the scientific understanding of the components within biodiversity. With this report, and its shorter companion piece, *Global Biodiversity Assessment for Policy-Makers,* WRI demonstrated that the root causes of biodiversity loss lie in the manner in which people use resources. The *Assessment* identified changes in human attitudes to nature as the single most important factor in enhancing biodiversity protection. It advocated educating communities about the "full social costs of their actions while enabling investors in conservation to reap the benefits."[64]

All this translates into some notable examples of effectiveness. When environmental NGOs employ participatory strategies such as grassroots networking and community education, they can be effective in achieving the ecological-economic linkage. These participatory strategies are particularly effective in defining issues within the proper ecological-economic context, at least on the local level, and then mobilizing communities toward that definition. Participatory strategies complement mainstream strategies in this sense because they assuage, to some degree, fundamental definitional problems that surround sustainable development on the local level. Where participatory strategies fall short, though, is in their ability to access decision-making points and implement programs. At times, as we have seen, NGOs are effective here but only when participatory strategies are accompanied by more mainstream approaches. To better understand this symbiosis, let us turn to the third and final linkage, that between short- and long-term considerations.

Short-Term–Long-Term Linkages

Our last linkage is perhaps the simplest of all. It rests on the assertion that what one does today affects the options available down the road. Unfortunately, the structure of politics creates incentives for those who operate counter to this intuition. Politicians play a political game that rewards those who produce results in the present. It rarely pays dividends for the future. It simply is not possible to get elected for what happens years later. This rationale, however, assumes a false choice. It assumes that an unbridgeable gulf exists between what is good for those of today's generation and what is good for those of tomorrow's generation. Much as with the false choices presented in the tensions between ecological and economic concerns, NGOs provide a promising medium for negotiating the problems found here. Tensions between interests of the present and interests of the future do exist, but they are not untenable.

While not an NGO itself, GreenCOM, a division of Academy for Educational Development, provides valuable assistance to NGOs and highlights the extent to which participatory strategies such as community education can be used to establish links between short- and long-term considerations. GreenCOM publishes a variety of documents for use in the field and a series of helpful discussion papers.[65] Acting on a grant from United States Agency for International Development (USAID), GreenCOM frames ecological improvement, including habitat conservation for biological diversity, in terms of behavior changes.[66] Groups must educate local communities, both here in the United States and the more commonly cited individuals in developing states, if they are to be effective in biodiversity protection. People need incentives to act in a sustainable manner; to provide those incentives, "you need to beat the drum all the time," says Peter Spain, administrative director of GreenCOM.[67] NGOs often beat this drum through the participatory strategies of grassroots networking and community education. It is here that they provide both immediate and long-term assistance toward an issue.

Earthwatch Institute incorporates this line of reasoning, stressing the need to connect short- and long-term considerations through education. A variety of teaching aides are available at their Web site, including resources and activity projects for students in Kindergarten through twelfth grade. Just a few examples of profiled programs are Ecuador Forest Birds, which investigates the ecology of birds in order to help protect those birds and their shrinking tropical forest habitat; Costa Rica's Dry Forest, which explores insects' role in maintaining a threatened habitat; and Bolivia's Savannah Birds, which monitors the effects of forest fragmentation and burning.[68] Like many groups in the late 1990s, Earthwatch targets children because they are both a visible symbol and practical instrument for the long-term perspective. Defenders of Wildlife tackles this same objective in its sponsorship of Project Wild. An interdisciplinary conservation and environmental education program for children in kindergarten through twelfth grade, Project Wild provides an array of teacher lesson plans as well as hands-on activities that enhance students' ability to master basic concepts. Even the business-like Environmental Defense has an education section complete with a teacher's guide for discussing various topics.[69]

These all deserve praise. Yet what makes Earthwatch unique, and perhaps more effective than most in establishing the short- and long-term linkage, is its consistent emphasis on education throughout the population. As discussed in the earlier section examining educational initiatives as a strategy, the objective of educating the public permeates all the scientific research that Earthwatch sponsors. Earthwatch, in conjunction with the Dodge Foundation, also funds teachers who wish to work in the rainforest for two- to three-week stints, and thus enhance their curriculum. Every year, more than four hundred teachers throughout the world directly participate in Earthwatch research thanks to educational fellowships that the NGO awards. By the end of 2003, a total 3,875 teachers had taken advantage of this opportunity. The obvious educational implications here center on the creation of a multiplier effect where teachers influence students, who in turn get more involved and perhaps even become scientists themselves. Such programs also relate to the domestic-international linkage in that they create partnerships between scientists, educators, and the general public in hosting sites and home countries.

Volunteer feedback at Earthwatch Institute underscores this point. Individuals have high praise for their Earthwatch expeditions, noting that it is wonderful to know that all discoveries are not already made and aired on PBS. Using volunteers has also shaped the actual practice of research, particularly influencing a philosophical change that treats public awareness as instrumental to conserving biodiversity. According to Gabor Lovei, a biologist from Hungary who worked on a 1999 project in New Zealand, "[Science] has become more interactive, has started to influence the way questions are asked, and the way science is perceived and presented."[70] The volunteers have come to actually shape the questions the research asks as well as how the research is conducted. The volunteers then have both a short- and a long-term impact. They routinely state that their Earthwatch expedition is an experience they will recount to their grandchildren, that they want to share their experiences in being a part of gaining new knowledge not only with those back home but those of generations to come. Earthwatch builds this into its participatory strategic agenda. Earthwatch demonstrates that, like the Sierra Club, it believes that people are the real

source of power by the prominent display on its Education Programs Web
page of a powerful quotation by Elizabeth Dowdesnell, executive director
of the United Nations Environment Program:

> We realize that there are only two sources of real power: people and
> money. Since the environmental cause will never be able to muster the
> kind of money needed to deal with existing and emerging problems we
> should focus on building people power. The future of the environmental
> movement rests on a culture of global environmental citizenship, where
> each and every one of us understands his or her responsibilities.[71]

When targeting people themselves, NGOs are also targeting the short-
to long-term linkage. World Resources Institute recognizes this and hopes
to build biodiversity awareness in primary and secondary schools with
precisely this objective in mind. On their 2003 Webby-nominated Web
site, for example, WRI recommends five specific points for teachers to
incorporate so that this linkage can be achieved.[72] These suggestions—
emphasizing biodiversity conservation efforts close to home and discussion of medicinal and agricultural contributions of biodiversity—clearly
contribute to each of the other two linkages.[73] Still, the underlying objective is to draw connections between decisions made today and ramifications felt years from now. Utilizing young students as a medium takes expressed advantage of the long-term perspective that our third and final
linkage requires.

One of WRI's strategic partners, World Wildlife Fund, takes a similar approach. As it notes in its publication *World Wildlife Fund at Work,* WWF
utilizes public education efforts with the objective of spurring the general
public to "take action before it is too late." Its "Windows on the Wild" environmental education program, for instance, provides workshops, activity
guides, and other materials that target middle-school students. WWF seeks
to engage young people because WWF believes these children might help
influence their parents today and tomorrow. But more importantly, WWF
sees these programs as a long-term investment, one where progress is
measured in years instead of days. WWF believes that "fostering a broader,
more active constituency . . . by teaching young people about the importance of biodiversity and the effects of global economic and social forces
on biodiversity" is one of the soundest strategies of all.[74] These efforts extend beyond the United States, as WWF also recognizes the need for
domestic-international linkages in the strategic approach of education initiatives. WWF incorporates environmental education throughout its field
projects in Africa, Asia, and Latin America.

Similarly, an array of workshops, environmental journalist seminars, and seminars for educators combine to reinforce educational awareness as a top priority for Conservation International. CI even sponsors an environmental journalism contest in Guyana, to encourage journalists to publish more high-quality articles about the environment. Their expressed purpose is to enhance public awareness of local conservation issues, and, thus, improve the decision making of governmental officials. CI seeks to link short- and long-term considerations. Of course, these efforts could easily be seen as condescending, even imperialistic, if they were not combined with educational efforts aimed at CI itself. Conservation International attempts to head off such criticisms by getting to know the local communities in which it works. And, it bears noting again, many of those on the ground in CI are themselves citizens of that country. Staffers sincerely believe that understanding the needs of those who live in an area, both their development and conservation concerns, is critical to any long-term conservation program.

With its origins on the coast of Texas in 1986, Ocean Conservancy's aforementioned International Coastal Cleanup provides one final example of how environmental NGOs seek to connect short- and long-term considerations. This cleanup relates to biodiversity in that more than 50 percent of all marine debris is some form of plastic, which is often mistaken as food by marine animals. When ingested by these animals, plastics cause serious internal injuries, including intestinal blockage and starvation. Plastic materials such as the ubiquitous six-pack holder rings and fishing lines also entrap and kill thousands of whales, sea turtles, seals, and birds each year. With over half a million volunteers in over ninety different states, this annual, one-day event was designed not only to remove debris; the Ocean Conservancy also uses the clean-up effort to collect data on amounts and types of debris, to educate the public about this issue, and then combines this information and informed public to affect change at all levels, from the individual to the international. The true measure of effectiveness here is not how much trash volunteers collect. It is what happens long after the final containers of garbage are carried away. Ocean Conservancy uses its cleanup to spawn recycling programs, adopt-a-beach programs, stormwater system overhauls, and general public education. It hopes to create a "ripple effect in all directions—jolting people awake to the magnitude of the problem, triggering new ideas for workable solutions, spreading interest, enthusiasm, and dedication . . . [until] these ripples become great waves of change."[75] The organization seeks to connect people and their immediate actions to the long term. Undeniably, education is a crucial component to achieving long-term trash reduction.

Despite these examples of effectiveness, though, NGO attempts to make the short-term–long-term linkage meet with mixed results. At times, NGOs have been able to define the issue in the proper context. At times, NGOs have accessed effectively the necessary decision-making points to relay this definition. And at times, they have been able to mobilize a group or even an entire community, applying both short- and long-term thinking in implementing a program. Still, NGOs have not been able to accomplish all four of these points consistently. There is much work still to be done on this linkage. NGOs must learn to incorporate an even longer timeframe in the initiatives that they pursue. This can be difficult given the short-term demands prevalent in our society. Limitations to the very transmission of information itself illustrate the difficulties that NGOs encounter as they seek to attract support for long-term problems. Fordham University's Everette Dennis, for example, identifies a short attention span within the general public as the dominant factor guiding limited media exposure of long-term problems. Continuity of topic coverage is all but impossible if people "can handle only three topics at any one time."[76] NGOs must continue the push to abandon traditional practices of jumping from brush fire to brush fire regarding environmental problems.[77] Too often, it has only been a crisis event that generates the attention levels needed to mobilize support for major policy issues, environmental or otherwise.

This cannot be the case with biodiversity loss. If the planet reaches a stage that is clearly a crisis, the phenomenon of irreversibility prevents adequate resolution. Once we realize that a crisis is at hand, it may very well be too late. No action, no matter how drastic, may be enough to reverse the process.[78] Thus, in the absence of other solutions, the participatory strategies of NGOs arise as possibly the last real hope for forcing decision makers to enact policies that consider the ramifications of the decisions they make today. No ceiling need be set on imaginative initiatives here. In fact, astronauts such as Spacelab veteran Tom Lind provide an intriguing commentary of note. As Lind states:

> This solemn planetary awareness may well be the most important legacy of the space age. It seems obvious to us astronauts that if we could get the leaders of our various nations to meet for peace conferences and other international negotiations in space instead of Geneva, they would reach considerably different and one hopes more appropriate solutions to our problems.[79]

No one is seriously proposing negotiations in space, of course. But the potential psychological effects from this imagery are useful.[80] And NGOs

Table 10. NGOs' Participatory Strategic Contributions to Three Linkages

	Domestic-international linkage	Ecological-economic linkage	Short-term–long-term linkage
Grassroots networking	Ocean Conservancy Sierra Club WWF	**WWF** **CI** **Environmental Defense**	Ocean Conservancy Earthwatch WRI WWF CI
Community education	WWF Ocean Conservancy Earthwatch TNC CI	**WWF** **CI** **WRI**	Defenders Earthwatch WRI WWF CI

could well exploit this imagery much as they have utilized the Apollo space photographs in further establishing the third and final linkage between short- and long-term considerations. Its simplicity does not make it any less important than the initial two linkages.

Summary and Findings

Participatory strategies add much to the effectiveness of NGO strategies in completing the three requisite linkages for transnational biodiversity protection. Table 10 shows those NGOs that utilize the two participatory strategies expressly for this purpose.

Table 10 shows that environmental NGOs can be remarkably effective with the participatory strategies of grassroots networking and community education as they attempt to solidify the second two linkages, those between ecological and economic considerations and short- and long-term considerations. While we have discussed various gaps that remain in these two linkages, excellent examples of effectiveness do exist on the local level. It is only when transposing these regional cases to the global arena that difficulties arise, as seen in the inability of any of the environmental NGOs to apply effectively participatory strategies to achieving the domestic-international link. Thus, as with mainstream strategies, participatory strategies to date have been unable to draw this fundamental connection. Turning to a summary of the conditions for effectiveness helps explain this further. Effectiveness here, as well as with the other two linkage hypothe-

Table 11. Effectiveness of Participatory Strategies in Establishing Three Linkages

Domestic-international linkage	Ecological-economic linkage	Short-term–long-term linkage
Improves the ability of NGOs to define issue and take advantage of access points but unable to date to mobilize or implement effectively.	Better at defining the issue than mainstream strategies, although this is still problematic globally; also able to mobilize but weak on accessing decision-making points and implementation.	Able to meet all four indicators at times, but again not consistently.

ses, relies on the ability of an NGO to meet each of the four criteria outlined first in chapter 1 and further discussed at the end of each linkage section in this chapter. Table 11 summarizes the status of the effectiveness of participatory NGO strategies toward establishing the three fundamental linkage hypotheses.

What seems to be increasingly clear, then, is that effective transnational biodiversity protection depends on the ability of environmental NGOs to take advantage of the power within unique combinations of the above linkages by integrating a number of distinct strategies. Those groups that are able to perform these tasks exhibit higher degrees of effectiveness. Earthwatch's participatory emphasis on scientific research as explored above is an excellent example. It also demonstrates the degree to which, in practice, the division between mainstream and participatory strategies is nowhere as clear-cut as it appears on paper. While some organizations obviously perceive grassroots lobbying to be an important first step along a linear evolution toward progressively more sophisticated tactics, other organizations maintain that any effective strategy must maintain grassroots efforts throughout. Certainly participatory strategies in the shape of grassroots networking and education initiatives add layers of much-needed activism to any campaign. They generate a critical mass by capturing the publicity needed to jump-start most campaigns. Still, to reach maturity, the environmental movement must find the proper balance between mainstream and participatory strategies. Grassroots first generated the support needed to create a mass movement, and some retention of these roots will always be necessary. Without it, as Conservation International's Cyril Kormos states, the entire movement is in danger of being co-opted by whatever interests are most powerful. Yet the extent to which participatory strategies are needed in more mature campaigns remains remarkably understudied.

To be most effective, should an organization concentrate its efforts first on mainstream efforts before turning to participatory ones? Or inversely, should an NGO emphasize first participatory and then mainstream activities? The answer in both inquiries is no. The relationship is more complicated than a simple linear interpretation. Participatory and mainstream strategies intertwine with one another throughout a protection or policy initiative. A simple linear progression from one strategy sub-grouping to the other does not exist. Participatory and mainstream strategies, while semi-distinct, also buttress each other. Each enhances the other's potential effectiveness. And the most effective linkages to date occur thanks to NGOs utilizing both such approaches, either by themselves or in partnership with other organizations.

Participatory strategies, thus, involve much more than just noisemaking. The specific strategy of grassroots networking provides the three fundamental components for any environmental strategist as mentioned earlier: money, media, and people. This last prerequisite cannot be understated, and it is the fundamental organizing principle for the Sierra Club. As Dan Seligman of the Sierra Club explains, without a mobilized force, no social change has ever been enacted in the United States. The Sierra Club thus supports citizenship at every level. It believes only active and environmentally conscious citizenship can shepherd to completion an issue as complex as biodiversity protection. Other groups such as World Wildlife Fund concur with this general assessment, stressing the need for capacity building as an integral aspect of citizenship.

These recommendations must be couched with the warning that limits to what one NGO can achieve on its own do exist. Participatory strategies actually underscore this point as they highlight the need within the environmental NGO community for heightened awareness about the strategic value added of each program an NGO implements. The most effective groups develop a strategic niche instead of simply chasing the hot issues of the day. This approach does not deny the awesome potential inherent within various combinations of strategic approaches. It simply means that one NGO should not attempt to employ more than it can handle. When organizations such as Earthwatch consciously limit themselves to dual efforts such as their scientific research and community education, effectiveness is clearly enhanced. The ability of an NGO to contribute to the work of other NGOs is improved as well.

Participatory strategies do not deny the power of mainstream approaches, but instead add valuable layers of support. Indeed, the relationship between participatory and mainstream strategies could best be de-

scribed in symbiotic terms. Grassroots networking and community education initiatives both buttress mainstream strategies and, in turn, require assistance from these same mainstream approaches if they themselves are to
realize their full potential. That is, participatory and mainstream approaches
are most effective when used in conjunction with one another. The need
for such cumulative approaches need not be concentrated within one organization, although at times this certainly simplifies the process. Organizations can instead team up, either explicitly or implicitly, in their efforts.

Coordination pays dividends on a higher plane as well. When considering the larger picture of biodiversity protection generally, truly effective
policymaking is only possible when NGOs from all political stripes not only
make active contributions, but also remain cognizant of what others within
the environmental NGO community offer. This type of strategic analysis allows for the synergistic properties discussed at the end of this chapter to
arise. It makes the efforts of the environmental NGO community total
more than merely the sum of individual (and at times divergent) parts.
Within this context, the discussion in chapter 4 turns to the organizational
structure of environmental NGOs. Only by understanding the constraints
and benefits of the specific structural design within NGOs can their synergies be maximized. Examining organizational structure also uncovers the
instrumental biases inherent to the tools of the NGO trade and thus enhances future carpentry efforts in building the next ark. Complete understanding of neither participatory nor mainstream strategies is possible
without such analysis. Much as comparative politics scholars such as David
Elkins and Richard Simeon have argued about the impact of culture upon
civic involvement, organizational structure within NGOs, while not the
final determinant of strategic effectiveness, certainly determines the available options for NGOs to utilize.[81] Effectiveness in achieving the three fundamental linkages, whether a group applies mainstream or participatory
strategies, is dependent upon these intervening organizational characteristics. NGOs should work in this arena as well then.

Working on themselves:
Improving organizational structure

INTRODUCTION

Sometimes change really is revolutionary. Sometimes it takes a palace revolt. That is how Conservation International was born. In the late 1970s,
two men at The Nature Conservancy decided it was time to expand their
NGO's domestic focus. Spencer Beebe and Peter Seligman started up an
international program at TNC. Seligman, who at the time had been working out of California, took a sabbatical, traveling down to Latin America
to learn Spanish. That trip convinced him that, if they really wanted to save
species, TNC had to expand south of the U.S. border. Not many groups
were truly transnational at the time, though, and TNC was trying, in a way,
to create the wheel. It certainly experienced its share of growing pains.
The organization's philosophy then was that the best way to save nature is
to own it, to put up a fence and keep people out. This approach had worked
remarkably well in the United States to this point, but TNC soon discov-
ered that same approach did not work as well in the tropics. Latin America has much different land tenure issues. People live more directly on the
land, needing to access it daily. Tree-hugging gringos coming down and
taking away this land, this source of livelihood, often made matters worse.

These frustrations were further magnified internally within TNC. The
international staff believed that their division lacked appropriate autonomy,
that the domestic program occupied too much of the overall budget.[1] When
these internal disagreements spilled over into the firing of Peter Seligman,
thirty-four staff and a handful of members from the board of directors
resigned as well. By late January 1987, they were "plotting" a new road—
and a new NGO, Conservation International.[2] Within the first couple weeks,
Conservation International lined up donations totaling a million dollars
and within the first few months launched the first ever debt-for-nature
swap. But some acrimony remained between CI and its maternal NGO,

TNC, at least for the first year. Lawsuits were filed. Legal fights over who would own the files erupted. At one point, TNC actually locked its doors and put chains on them. The atmosphere was tense, and the story continues to live within the environmental community as practically an urban legend. Even to this day staffers at both TNC and CI speak in only general terms about the original disagreement, despite the fact that only three individuals from the original "revolution" remain at CI today.

This is both good and bad. It is a positive development in that both organizations now realize that there is plenty of room to operate. There is a lot of biodiversity to share. And both organizations should be applauded for not only recognizing this but also partnering with one another directly in a number of regions around the world where they both work. But the lesson behind the split must be noted as well. NGOs must continually remind themselves that there are needs for different types of groups. Both TNC and CI have grown enormously since the split, and both have enjoyed tremendous successes. Their story, including past differences, is not something of which to be ashamed. One need not dwell on it, emphasizing the negative. But neither should these differences be glossed over or hushed up.

Over time, a number of other NGOs have spawned from similar circumstances. Most of these have been velvet divorces. During the "second wave" at Conservation International, for example, ten staffers, including current CI president Russ Mittermeier, moved from WWF to CI in July 1989. Similarly, Earthjustice distanced itself from its parent, the Sierra Club, by changing its name from Sierra Club Legal Defense Fund in 1997. Earthjustice has always been legally independent with its own staff, budget, and contributors, despite close ties to the Sierra Club, one of its most frequent clients. The Sierra Club is not Earthjustice's only client, though, and the group renamed itself in 1997 to better describe its relationship with other environmental organizations. But in this case, fundamental strategic differences and a degree of inner politics were also involved. When then Sierra Club Legal Defense Fund petitioned for listing the spotted owl as endangered in the late 1980s, for example, the Sierra Club's executive director Michael Fischer worried about public backlash for his organization, and even hinted that perhaps a name change should be in order. NGOs must not forget that these differences are natural. They will inevitably develop, and they will, more often than not, be for the best.

As such, NGOs must continually examine themselves. The task NGOs face in terms of working on themselves is fairly basic, but paradoxically also laden with difficulty. Peering into the looking-glass self can be difficult. You might not always like what you see. But environmental NGOs generally understand this—and the basic fact that only by doing so can they continue to improve their ability to protect biodiversity effectively. Some degree of change is to be expected, even welcomed, given the issue area of focus. Those who seek to protect biodiversity undergo their own sort of speciation. They evolve to better serve the needs present. And it is their NGO status, unlike more bulky bureaucratic state agencies, that fosters these developments. Nevertheless, too much change with new NGOs and new strategies draws groups away from their expertise. And NGOs must keep this in mind as they continually evolve.

ORGANIZATIONAL CHARACTERISTICS

Connecting the three fundamental linkages of domestic and international, ecological and economic, and short- and long-term considerations as they relate to transnational biodiversity protection is an arduous task. NGOs using both mainstream and participatory strategies encounter daunting obstacles, as we have already seen. At times, together, they are able to overcome these hurdles to effectively establish the three fundamental linkages to transnational biodiversity protection. At times they are not. But NGO strategies are never applied in a vacuum, as evidenced by the fact that no one strategy or even combination of strategies is effective all the time. Strategic effectiveness is dependent on other factors besides simply the strengths and weaknesses of a specific strategy. Organizational characteristics often intervene, presenting both considerable constraints and much-needed supports for specific strategic initiatives. Indeed, the linkages discussed in the previous two chapters would not be possible without the nuts and bolts of organizational structure that this chapter addresses. Organizational characteristics influence not only the effectiveness of a particular NGO strategy but also the effectiveness of the three fundamental linkages required for transnational biodiversity protection.

The organizational characteristics of NGOs, however, are often neglected when scholars examine transnational biodiversity protection. NGOs have proven to be an effective medium for instituting transnational biodiversity protection in several cases, but, to take full advantage of the promise they offer, we (including the NGOs themselves) must come to a better understanding of how these organizational characteristics affect the objectives that NGOs pursue. This chapter recognizes this level of significance, that organizational characteristics are analogous to the carpentry training and tool maintenance needed to build the next ark. Only properly trained and equipped NGOs may effectively utilize their strategic tools in constructing the next ark, in establishing the requisite linkages for transnational biodiversity protection. Furthermore, these linkages are only possible if the strategic tools that an NGO employs are adequately maintained. Specific organizational characteristics can either provide this proper level of maintenance, or they can allow the strategic tools of NGOs to fall into disrepair. They can support strategic initiatives, or they can constrain them. NGOs must do more than work within the system and work with people. They must work on themselves.

To be able to use the tools of the biodiversity protection trade, namely the mainstream and participatory strategies outlined in chapters 2 and 3, NGOs require essentially the very conditions that political scientist David Truman identified in the early 1950s for governmental processes more generally. They require adequate cohesion, internal leadership, and basic resources—both of the personnel and financial dimension.[3] From this prescription one may cull five key organizational characteristics within environmental NGOs. They are the following:

- general demographics (including both legitimacy and resiliency in age, funding, membership, and staff)
- decision-making style (including centralization, transparency, and imagination)
- willingness to engage in partnerships
- recognition of appropriate targeted constituency
- internal support for specific strategic concentration

Like the strategies discussed before them, considerable overlap exists here. These are not entirely discrete characteristics. Still, each in its own right acts as a critical determinant in the effectiveness of NGO attempts to make the fundamental linkages required for transnational biodiversity protection.

General Demographics

The broad category of demographic composition includes the most basic organizational characteristics, typically the data displayed in the annual reports issued by NGOs themselves. Age, funding, membership, and staff are all included within this characteristic. The greater each of these values becomes, the greater the level of support for strategic effectiveness. Older, wealthier, and larger (both in staff and membership) NGOs typically enjoy basic resource advantages over younger, poorer, and smaller NGOs. These resource advantages translate into greater legitimacy and resiliency for an NGO. They support NGO strategies, making them more effective. Let us look at these in a bit more depth.

Resource advantages begin with the ability to expend money. Those groups with a substantial operating budget have a distinct advantage in certain strategic initiatives. Financial viability is undeniably an integral component of effectiveness, particularly for those NGOs that stress costly mainstream strategies such as lobbying and litigation—although significant capital is often needed in even the most decentralized participatory programs. Funding is also important as it relates to the staff of an organization for obvious reasons. As with any group inside the Beltway, opportunities continually arise for Washington, D.C., staff members to jump to another position. Those NGOs that can afford to compete with other employers salary-wise often enjoy a greater degree of stability than those that cannot.

Perhaps the most overlooked variables when it comes to legitimacy and resiliency are those of the staff and membership itself. People do matter when it comes to NGO effectiveness. Seema Paul, senior program officer for biodiversity at United Nations Foundation, the UN entity established thanks to a gift from CNN founder Ted Turner, believes that this cannot be overstated. "It's the personalities involved that have made all the difference. It's up to people . . . [as] a lot of it is the human element."[4] The sheer number of individuals at an NGO's disposal matters in both the formation stages of an issue and the implementation phase for a project. In qualitative terms, moreover, staff reputation can be the determining factor, from shaping the priorities of a member of the U.S. Congress to convincing fellow NGOs across the world to support a specific proposal within international forums. This human element is also an attribute stressed by Mick Seidl, chief program officer for environment at the Gordon and Betty Moore Foundation. Established by Intel founder Gordon Moore in 2001, this granting institution created quite a buzz in the environmental community when it handed CI a $261.2 million grant in October of that year. As Seidl asserts, "In the end, its all about a specific group of people that are very dedicated."[5] Seidl was speaking about the NGO staff, but it is important to remember that this focus on people is true on two dimensions. Both those within the organization and those who support it from outside play a role in shaping demographics. The Sierra Club, for one, prizes its membership as active citizenry devoted to harnessing the benefits of political power.

Access points from foundations to the United Nations apparatus to U.S. government agencies identify the characteristic of staff prestige as an under-acknowledged and undervalued one within the NGO community. This significance becomes all the more important when a disturbingly high turnover rate is exposed within environmental NGOs. Exact numbers unfortunately are not available here, but numerous conversations with NGO officials, often off the record, indicate that this is a recognized problem within environmental NGOs. Often this turnover involves moving from one environmental NGO to another, such as when Dan Seligman moved from World Resources Institute to the Sierra Club after a two-year stint at WRI. Sometimes staff shifting occurs from NGO to NGO but not necessarily in an environmental context. For example, Estraleta Fitzhugh came to World Wildlife Fund in January 1998, after working at Amnesty International in a similar governmental relations position. And still other times, NGO staffers move to governmental positions or the more lucrative for-profit sector.

While he does not believe this to be a serious problem, the Moore Foundation's Seidl identifies precisely this phenomenon. "NGOs are stealing people from each other on a regular basis," in his view. While endemic to Washington, D.C., generally, this is a particularly severe handicap in the environmental community. High rates of personnel turnover limit not only institutional memory but also the maintenance of the social capital that NGOs, whether using participatory or mainstream strategies, seek to exploit. And it violates the basic precept that political scientist William Browne identifies for interest groups, that people themselves are an integral component of any organized groups' efforts to influence public policy and the public policy-making process.[6] There is one possible positive spin to this turnover, however, for it shows that groups are often aware of one another on the most individual of levels. As long as the split is amicable, moreover, turnover to another NGO facilitates future cooperation. Some institutional memory is lost, but benefits can outweigh these costs.

Decision-making Style

Decision-making style incorporates the concept of cohesion that Truman identified as a prerequisite to the smooth transaction of organizational business. As the key to all-important standard operating procedures for an organization, it represents the internal rules by which an organization governs itself. It is the organizational culture. The dimensions to decision making of note here include centralization, transparency, and imagination. Hierarchical decision making is efficient but decidedly non-participatory. Few access points exist and a monoculture of ideas often develop as the same individuals in upper management are responsible for new initiatives day after day. Change can only be instituted from the top down. Efficiency prevails but at the expense of imagination.

Decentralized structures, on the other hand, foster participation and the belief that multiple access points enhance the opportunity for meaningful change. Opening up the channels of participation not only extends the political power in an organization to previously neglected entities. Decentralization also enables the actual membership of a group to have input into the decision-making process, a characteristic that American University's Ronald Shaiko believes critical to effective political influence. From local to global, Shaiko asserts organization leaders best serve their constituents by "acting with their members in a more concerted effort."[7] Without these steps, effectiveness is limited. Unfortunately, this decentralization also brings its share of costs to the table. Chief among these is the fact that decentralization extends the time needed to make decisions—at times to the detriment of both an organization generally and the solving of the actual issue more specifically. Spreading authority around in this manner, much like the system of separated powers in the United States government, makes change slow and painful. While disasters are thus avoided, so is the potential for the needed paradigm shift.

Both transparency and imagination are also variable within the decision-making process, with implications for the effectiveness of strategic initiatives. Transparency denotes the openness of an organization, a particularly fruitful trait if one hopes to establish public acceptance. Its utility is well established within the context of improving treaty compliance, for example.[8] Groups that open up their decision-making channels for scrutiny by not only their staff but also their general membership and even the public at large clearly enjoy long-term benefits in terms of enhanced reputation and increased social capital.

Imagination refers to the ability of an NGO to incorporate a diverse set of tactics within its cache of strategic initiatives. Being too conservative and relying on one tactic or one access point can become a handicap for NGO decision-making. NGOs must not try to do too much, though, as there is a real danger in over-extension.

Partnerships

The willingness of NGOs to establish and maintain partnerships is also an emerging characteristic of importance. The term "partnerships" includes broad coalitions and networks with business or governmental agencies as well as specific initiatives between two or more NGOs, either of the large international variety or smaller, local groups on the ground. The utilization of partnerships, of course, could certainly be categorized as a strategy in its own right. But partnerships are most notable for the impact they have on other existing strategies in an NGO's repertoire. NGOs could and do emphasize the strategy of partnering with other groups, agencies, and businesses, but they need one of the more discrete (albeit by no means entirely discrete) strategies discussed in the previous two chapters as well. Networking may be either formal or informal, as UN scholars Leon Gordenker and Thomas G. Weiss point out.[9] The mere process of creating bonds enhances the forum for discussion, improving efforts at coordination and collaboration. Whether an NGO stresses formal structures and mainstream approaches or interpersonal relationships and participatory strategies, the organizational characteristic of willingness to utilize partnerships enhances effectiveness. Networks and coalitions allow NGOs to "exert an influence above and beyond their weak formal status."[10]

Coalitions, especially large coalitions, also may create an aura of urgency and enable NGOs to jump-start a previously dead issue. Thus, while coalitions often do prevent outright conflict from arising within the environmental community itself, as will be noted in the later sections of this chapter, more is at work here than merely minimizing conflict. Partnerships improve the specific strategic initiative an NGO chooses to emphasize by allowing the NGO to supplement its efforts with the strategic initiatives of others. As is increasingly evident in environmental circles, partnerships represent the explicit recognition that some groups perform certain tasks better than others do. They are a concerted effort to maximize efficiency and effectiveness.

Coalition building, though, places demands on both energy and time. It also requires resources. NGOs that channel their efforts into partnerships must sacrifice time, energy, and even funds that may have gone to other efforts. And partnerships are not a panacea. They will not cure everything, sometimes even creating more problems than they solve. As UNF's Seema Paul notes, there are costs in terms of efficiency every time you employ partnerships. "There are time constraints and delays. Things take a lot longer, and the question is whether biodiversity can afford those costs. Will actions be fast enough? The moment you have two or more partners, it takes much longer to get results achieved."[11]

NGOs must also consider the negative ramifications of partnerships, both when initiatives go sour and possibly even when they are fruitful. Failed partnerships may exacerbate an already tenuous relationship. Failed partnerships may renew incentives to compete against one another. Failed partnerships may also spur an unhealthy competition for resources, both in the human resource of membership and the economic resource of financial capital. Indeed, groups are reluctant to share donor lists, according to the Moore Foundation's Mick Seidl, for this very reason. But it is precisely this data sharing, this transparency, that is needed. In rare cases, moreover, even successful partnerships may be partially destructive. They may attach the unwanted baggage of another group to an NGO. And they may dilute the sweetness of success simply by demanding division of the spoils of victory. Any time rewards are shared, there is the potential for acrimony to develop—whether due to unwarranted jealousy or legitimate complaints about relative reward distribution. Despite these potential negative ramifications, this analysis shows that the benefits of partnerships far outweigh their costs. Partnerships clearly support the entire spectrum of strategies NGOs utilize.

Targeted Constituency

The ability to form partnerships also enables environmental NGOs to better address a fourth characteristic that can be constraining when it fails to recognize the transnational nature of biodiversity protection. Too often, the organizational structure of an environmental NGO is unable to address the myriad of different constituencies needed to negotiate an issue as complex as transnational biodiversity protection. Partnerships alleviate this deficiency. They allow NGOs to work at different levels and in different sectors. In short, they allow NGOs to address different constituencies. This is a key point as it relates to the formation of fundamental public interests. As comparative political behavior scholar Russell Dalton contends in his analysis of public opinion in Western democracies, interest in environmental policy translates into public support for environmental protection.[12] Environmental NGOs must first generate general interest in transnational biodiversity, before they can mobilize support for its protection. At times, interest in major environmental issues erupts with a major event or catastrophe, but often environmental interests must be cultivated carefully. This is particularly the case with long-term issues that do not immediately victimize a specific group of people. Transnational biodiversity loss falls into this categorization. Some exceptions can be made here, such as when mammoth state development projects in the developing world (such as World Bank sponsored hydroelectric initiatives) eliminate not only species but also the homes and livelihoods of indigenous populations. Still, by and large, the effects of transnational biodiversity loss take time to develop, and they are spread over a variety of constituencies.

Ironically, then, what is actually *the* crisis event of the twenty-first century appears to be merely a slowly developing issue. It does not appear to be the crisis event upon which the public interest generally fixates. The analogy often used here is what is known as the "cooked frog problem."

When a frog is placed in a pot of boiling water, it quickly leaps out. The frog knows it's in trouble. But when a frog is placed in a pot of water at room temperature, he stays in the pot—even as the temperature is slowly ratcheted up to a boiling point. The frog does not recognize the changes around him, namely the warming of its environment, and slowly simmers to death.

Environmental NGOs must take steps to correct this erroneous image of biodiversity loss, this belief that the temperature of our pot of water is just right. If not, we may meet the fate of our friend the frog. This is true both figuratively in the context of the above "cooked frog problem" and literally in the context of the massive die-offs of amphibians throughout the world in the past two decades.[13] While community education highlights the variety of strategies that contribute to the tall task here, the organizational characteristic of targeted constituency operates as the defining element in any effort. Poorly targeted constituencies will constrain even the soundest strategy while targeting constituencies of a truly transnational nature offers much-needed support. Collective action in this case cannot be a spontaneous event, precipitated solely by a repressed, socially disadvantaged movement. Although there still is, and always will be, a place for collective action from these disaffected origins, issues such as transnational biodiversity protection demand that mobilization begin in other sectors as well. With the threat of irreversibility hanging over our collective heads, action born from a crisis situation is action that is too late. Protest politics must evolve, as Dalton explains, with better organization and further incorporation of educated, middle-class constituencies.[14] Even more, those in power or proximate to the seat of power must take a more active role. At least some of those with a vested interest in the system must be part of the change if there is to be any substantive change at all. Of course, the million-dollar question here is: How does an NGO facilitate this?

The short answer is that there are many viable options. These range from radical, monkey-wrenching activities that inflict financial damage on corporate entities to more conservative alliances where NGOs outline financial incentives for businesses going green. By targeting the appropriate constituencies, NGOs can execute this charge. They will expressly incorporate Dalton's discussion of political efficacy. Explicit in Dalton's framework is the adoption of the fundamental assumption that "one's political action can affect the political process."[15] As they target particular constituencies, environmental NGOs must incorporate this theme into their basic message. Emphasizing the efficacy of the NGO must be a fundamental component of their organizational structure. As Dalton has stated, political efficacy on the individual level is essential to any political action. "A feeling of political efficacy motivates individuals to become active in politics, while the absence of efficacy evokes political apathy and withdrawal. If one cannot affect the political process, why bother to try?" Dalton questions.[16] Environmental NGOs must instill this political efficacy on a group level—and expressly show how individuals can enhance this efficacy with their own participation. People have to be shown that they make a difference.

NGO constituencies in this context are much broader than official membership rosters. As William Browne contends, mobilization must involve more than just the "joiners." Members themselves are rarely sufficient targets. Environmental NGOs must target the public at large.[17] This is a difficult assignment. Too often the characteristics of constituent makeup and organizational objectives get lost in the shuffle of particular strategies and tactics. That is, organizations neglect their original goals of assisting their immediate constituency. While they may display mission objectives prominently on their Web pages and publish flourishing rhetoric in periodic membership mailings, NGOs constantly run the risk of losing focus upon their ultimate goal. This is especially true when NGOs are confronted with the daily demands of organizational maintenance.

NGOs in their position both above and below the state hold the unique ability to target multiple constituencies. At times, admittedly, this ability becomes more of a constraint than an asset as groups struggle to divide their resources in domestic and international programs—even to the extreme that it sparks outright mutiny within an organization, as was the case with TNC spawning CI. These internal difficulties, not surprisingly, affect the targeted audience as well. A confused message often emerges, one that fails to tie together the dependent strands of the equation. One constituency in the United States is conceptualized independently of a constituency in Costa Rica or Ghana or India, and valuable resources are lost. NGO strategies are only supported by this organizational characteristic, then, when NGOs explicitly recognize the constituencies that are being targeted. And to best exploit their targeted constituencies NGOs must better understand why people join environmental groups. They must better understand why people take an interest in an issue. Following the logic of collective action set out by economics scholar Mancur Olson in regard to interest groups, NGOs must better exploit the rational self-interest perspective if they are to ensure the characteristic of targeted constituency acts as a support rather than a constraint.[18]

Strategic Concentration

As detailed in chapters 2 and 3, the specific strategies an NGO applies in protecting transnational biodiversity are significant in terms of shaping its effectiveness. At certain points in time, under certain conditions, one strategy is better than another—in a specific locale. When attacking an issue with the breadth and longevity of transnational biodiversity protection, each of the strategies discussed makes its own contributions. No one strategy works all the time. All strategies make some contributions. Lobbying is more effective than monitoring during the initial agenda-formation stages. Yet without monitoring, as implementation proceeds, this initial lobbying would become practically meaningless. The two, like our subject, are interdependent. They are intrinsically tied together.

Paradoxically, until NGOs recognize that the interdependencies they seek to protect in the natural world also exist within their own environmental community, until they recognize the different strategic functions that various groups offer, discrete strategy applications will be ineffective. NGOs that orient their programs around not only the contributions they seek to make but also those that others seek to make (and are making) act in a supportive capacity. This entails avoiding the temptation that NGOs face of overextending their individual strategic initiatives. One NGO cannot perform all the necessary functions in the transnational biodiversity protection process. Transnational environmental groups must find their strategic niche with acute attention to the contributions that other organizations make. Time and again, NGOs are told by both members and foundations alike, "There are too many environmental groups. Why don't you folks work together?"[19] This is precisely the argument made by lobbyist scholar William Browne in his examination of U.S. interest groups. Browne contends that comprehension of policy niches is the key to long-term interest survival.[20] Strategic concentration on two or three strategies within the context of the contributions that other NGOs provide acts as a notable support. Disjointed applications of strategies without recognizing the role of other NGOs acts as a constraint.

Summary and Findings

The organizational structure of an NGO plays a key role in determining the effectiveness of the specific strategies that NGO selects in seeking to protect transnational biodiversity. Some characteristics support the requisite linkages in this issue area. Some constrain them.[21] Table 12 outlines both these supports and constraints:

Table 12. Organizational Characteristics as Supports and Constraints

	Supports	Constraints
General demographics	Finances, staff prestige, and a viable membership all aid an NGO in its various strategic initiatives as they enhance legitimacy and resiliency.	Personnel turnover limits NGOs more than they care to admit. Lack of resources and staff can also handicap an NGO.
Decision-making style	Transparency enhances the legitimacy of NGOs, and imagination is also a valuable asset. Centralization of decision making has both pluses and minuses as discussed under constraints to the right.	Centralized decision making often limits imagination, but decentralized decision making is less efficient.
Partnerships	Partnerships offer gains in economies of scale by sharing resources and creating a wider range of policy contacts. They also allow NGOs to focus on their own strategic niche as other groups fill in any missing gaps.	Partnerships cost time, energy, and money. They also force NGOs to share spoils of victory.
Constituency targeted	When able to emphasize the transnational nature of biodiversity protection, NGOs are then able to incorporate domestic constituencies from a host of states.	NGOs are often preoccupied with membership as the sole constituency. At times they mistakenly conceptualize biodiversity protection as an international issue only.
Strategic concentration	Synergies in two or three strategies display remarkable results, particularly when in combinations of mainstream and participatory varieties.	There is a danger of trying to engage in too many strategies at one time, thus spreading an NGO too thin and reducing overall effectiveness.

STRATEGIC SUPPORTS

Supports to Mainstream Strategies

Depending on their variation, the five organizational characteristics discussed above may act as either supports or constraints to the mainstream strategies seeking to make the three fundamental linkages of domestic-international,

ecological-economic, and short-term–long-term concerns. Strategic concentration, for one, can be a remarkable asset to NGOs when they limit
themselves to two or three specific strategies. World Wide Fund for Nature, the international affiliate of the U.S.-based World Wildlife Fund, demonstrates how this organizational characteristic supports strategic effectiveness as seen in its participation in the Ramsar Convention. Even before
the United States became a contracted party in 1987, WWF-International
combined research and monitoring to shape the international agenda of
Ramsar.[22] The Ramsar Convention requires parties to list one or more
wetlands as "Ramsar sites," use these sites efficiently considering multiple
interests, and supply information on the policies designed to implement
that wise use. NGOs such as WWF and Birdlife International along with the
NGO-IGO hybrid IUCN often serve as official advisors to the Standing
Committee that oversees operations of the Convention between its COP
meetings. NGOs, thus, serve in an official capacity and supply much information through the combination of research and monitoring strategies. Informal channels are also utilized as shadow lists compiled by NGOs influence the official Ramsar wetlands lists, for example.[23] What is particularly
notable here, aside from the obvious fact that NGOs and IGOs are working in official concert on this international treaty, is how NGOs actually
participate in what was traditionally only a state arena. They do so by exploiting the remarkable synergies of monitoring and scientific research.
The organizational characteristic of strategic concentration allows NGOs
like WWF to be remarkably effective in this case.

General demographics also contribute much-needed support, particularly when considering the actual people within an organization. The skills,
resources, and personal connections they lend to an NGO can add substantial support to strategies. Certainly absolute size makes some contributions
in the sense that a recognized name as one of the large conservation organizations may help staffers get into doors that might otherwise be closed
to them. Larger organizations have more resources at their disposal. As
Cyril Kormos of Conservation International states, "[This size] allows them
to add layers of analysis that others cannot. We provide a more accurate
presentation of the landscape."[24] Yet, while absolute numbers may at
times be necessary, they are clearly insufficient in and of themselves. Most
importantly, reputation of the organization and the specific individuals
within that organization give staffers an advantage in the strategies they
seek to initiate. Infrastructure in the form of pure numbers often opens
doors, but the prestige and respect that individual people establish determines whether that access will amount to actual influence. Knowing people

personally makes a difference, as Estraleta Fitzhugh, congressional liaison in the department of government relations at World Wildlife Fund, contends.[25] Networking pays dividends. Individual staff members can enhance strategic effectiveness.

The executive director of IUCN, Scott Hajost, agrees with this assessment. Top-quality staff are a valuable asset, he says. Political instincts and savvy can mean the difference between a strategy working and failing miserably. As Hajost states, to play the diplomatic game effectively, staff must know how government works, how international treaties work, and how to operate on the international stage. In the most basic terms, staff members best exploit their personal prestige by building coalitions—a trait to be explored shortly in terms of partnerships. The international prestige of Michael Oppenheimer, for instance, allowed him to recruit top scientists for Environmental Defense, Hajost contends.[26] Sam Johnston, program officer of Financial Resources and Instruments of the Secretariat for the Convention on Biological Diversity, echoes this sentiment. To Johnston, large NGOs like World Wildlife Fund do have an advantage in terms of boasting sizable staff, yet, he believes, effectiveness is ultimately shaped by individual personnel.[27]

Strength of personnel is further enhanced by imaginative, transparent, and decentralized decision making.[28] As Cyril Kormos explains, this is most often true when an NGO resists the urge to retain a tight hierarchy in the decision-making process. Conservation International attempts to incorporate this rationale when they host their annual planning meeting for a two-week period each March. This meeting draws staff from throughout the world to discuss plans for the next year. Over the course of two weeks, synergies evolve. Traditional, centralized decision-making paths are bypassed. The decentralized format, in turn, fosters creative thinking and measured risk-taking. It fosters the imagination component discussed earlier. As a dynamic organization, Kormos believes that CI is an exciting place to work. As a relatively decentralized organization, channels of communication are open. Even though chains of command clearly exist, younger staffers have the freedom to walk directly into CI director Russell Mittermeier's office at virtually any time. As Kormos notes, this is not true for all institutions, particularly those that focus heavily upon mainstream strategies.

The extent to which imaginative decision making acts as a supportive characteristic is also evident in the lobbying efforts of the Ocean Conservancy, which boasts the 1995 NGO of the year award presented by the National Oceanic and Atmospheric Administration within the Department of Commerce. As the Department of Commerce stated, the Ocean Conservancy was recognized for "its lasting effect on both public awareness and political action."[29] The source of this award is almost as important as the award itself in this example. Here an actual access point was recognizing Ocean Conservancy's work. This cannot but help Ocean Conservancy's stature in the Department of Commerce's eyes (as well as other U.S. governmental agencies and the public at large). It improves Ocean Conservancy's ability to lobby in the future. Ocean Conservancy has also received the President's Environment and Conservation Challenge Award, Keep America Beautiful Award, and three Take Pride America awards. Each of these further supports Ocean Conservancy's future efforts at lobbying— particularly given their relatively small staff size of approximately eighty-five to ninety individuals.

Probably the most consistent support mechanism among organizational characteristics, though, is the willingness of NGOs to engage in partnerships. Partnerships clearly enhance NGO effectiveness. Notable examples of those that do utilize partnerships indicate the potential this reserve holds. Environmental Defense boasts numerous such relationships, including the famous 1990 partnership with McDonald's to reduce overall paper usage and halt the use of foam plastic hamburger boxes. While this partnership was not directly targeting biodiversity protection, the overlap to our issue is obvious and twofold. For one, this program reduced CFC impacts on the global environment, impacts that contribute to global warming that, in turn, threatens species diversity. For another, it limited the need for paper products, the consumption of which requires habitat destruction. In many cases this is ecologically unique habitat that species diversity requires.

With this history of success, partnerships targeting more direct applications in the arena of transnational biodiversity protection are an emerging phenomenon. Environmental Defense also is paired up with the Ocean Conservancy, World Wildlife Federation, Greenpeace, and the National Wildlife Federation in protecting dolphins, for instance. The five environmental NGOs' joint testimony before the Subcommittee on Oceans and Fisheries in April 1996 increased the effectiveness of the mainstream strategy of lobbying in this case. Simply by virtue of a larger base of advocacy voices, this group statement appeared to be more legitimate in the eyes of those on the subcommittee. The fact that several different groups with divergent constituencies submitted a joint statement held greater sway than five separate, uncoordinated statements would have held, as Ocean Conservancy marine mammalogist Nina Young contends.[30]

World Wildlife Fund also provides a good example of how partnerships support mainstream strategies such as lobbying. Estraleta Fitzhugh spends the majority of her time engaged in lobbying bills the old-fashioned way. From briefings in committee meetings on the hill to individual meetings targeting key people regarding development assistance, she is intimately involved with the proposal stages of draft language for legislation. A critical component of her daily agenda is gathering support within the environmental community. Effectiveness, she explains, depends upon the extent to which WWF can get support from other environmental NGOs. By forming coalitions with other transnational conservation groups, WWF is more effective.[31] Earthwatch and the Sierra Club enjoy precisely this type of supportive relationship. Earthwatch periodically shares results of its various scientific expeditions with the Sierra Club. The Sierra Club, in turn, publishes these results in its own manuals and pamphlets. Given Earthwatch's status as a non-political organization that refrains from any lobbying activity in Washington, this partnership clearly enlarges their audience. It makes possible the third tier of their expressed mission: conservation.

The Save Our Environment Coalition is still another example of this organizational characteristic of partnerships at work. With fourteen of the largest environmental and preservation organizations in the United States, the Save Our Environment Coalition lobbies "to protect America's most special places."[32] The mainstream lobbying approach is magnified when fourteen highly visible NGOs sign on together. Six of the groups in this study are involved in the Save Our Environment Coalition: the Ocean Conservancy, Defenders of Wildlife, Earthjustice, Environmental Defense, Sierra Club, and World Wildlife Fund.[33] All six improve their lobbying strategy by working in concert with those of a similar mindset. Their widened advocacy base gives each NGO greater latitude and amplitude to apply in discussions with access points.

Defenders of Wildlife, similarly, seeks to widen its constituency, to improve its effectiveness through the use of partnerships. Rina Rodriguez, former international associate at Defenders of Wildlife, explains that, particularly in the international arena, working together in groups carries enormous benefits. Pointing to the lessons from NAFTA, Rodriguez believes division within the environmental community can be highly destructive. "Most of what I do involves working with other groups. International coalitions are constantly being formed."[34] These partnerships, furthermore, do not necessarily need to be with other environmental groups. For instance, Defenders had an agreement with the long-distance telephone company, Working Assets. In this agreement, Working Assets provided a toll-free number for activists to contact the American Farm Bureau Federation asking them to drop their lawsuit on reintroduction of wolves into Yellowstone National Park.[35]

The ability of Earthjustice to focus its efforts on litigation serves as still another example of the support that partnerships provide. Teaming with long-term partners such as Defenders, Environmental Defense, World Wildlife Fund, and, of course, the Sierra Club, also allows Earthjustice to meet another supportive characteristic, strategic concentration. Earthjustice especially owes a debt to the Sierra Club for allowing it to concentrate its resources on its primary strategy of litigation. The Washington, D.C., office of Earthjustice no longer has a media department per se. They do hire media consultants from time to time and have a director of communications in the national headquarters in San Francisco.[36] For assistance in Washington, though, Earthjustice often turns to the Sierra Club. But, by emphasizing litigation as their primary strategic initiative, by carving out a niche in this strategic concentration, Earthjustice is often able to return the partnership favor.

Indeed, many environmental NGOs reap rewards from partnering with Earthjustice. The Ocean Conservancy and the Turtle Island Restoration Network, for instance, both benefited from a partnership with Earthjustice in a series of legal victories preserving protections for the Pacific leatherback sea turtle. Earthjustice originally filed a suit on behalf of these two NGOs in early 1999, asking for stiffer regulations on fishermen who were endangering this species. In response, the Hawaii Longline Association challenged rules that would reduce the number of fishing days, limit the areas available to fishing, and require a federal observer on every boat. A mid-summer 1999 ruling in federal district court continued to support these regulations, thanks in large part to the continued diligence of Earthjustice—and its partnership with the Ocean Conservancy and the Turtle Island Restoration Network.[37] Earthjustice must resist the pressures to expand in the future so that this strategic niche of litigation is retained. While Earthjustice still focuses primarily on litigation as a strategy, their D.C. office has grown to include a policy and legislative vice president along with five to seven staff members who engage in periodic lobbying activity, which includes providing briefs for debate talking points on the House or Senate floor.[38]

Still another example of NGO partnership may be found in a map of global biodiversity priority areas that The Nature Conservancy created along with Conservation International, World Wildlife Fund, and Birdlife International. WRI, IUCN, and African Wildlife also participated in this exercise. Funded by the United Nations Foundation, this was the first time these groups shared their information on this level. By combining their data sets and specific program initiatives into one source, these NGOs gained a new appreciation for precisely how their individual contributions fit within the larger picture. The map partnership goes a step further, moreover, in that it makes gap-analysis possible. Not only do the organizations now understand who is doing what and where. They also have a better understanding of what is overlooked.

Unfortunately, more tangible applications of this map resource have been delayed due to political differences over which group's areas would receive prioritization. There is some nervousness among partnerships such as this because huge chunks of their specific programs become vulnerable. The intention of this project was to map out geographically where the various NGO programs overlapped. This benefit quickly was tempered by the fear that such overlap, in turn, suggested that remaining areas falling outside the overlap might not be as important. WWF, with its "Global 200" areas of biodiversity, incorporates more than the hotspots upon which CI focuses. WRI's frontier forests and Birdlife International's endemic bird areas, while displaying a degree of overlap, also exhibit disjuncture. Several NGOs are understandably nervous about potential cuts in their personal programs.

Partnerships are also beneficial to other groups that concentrate upon scientific/technical research or monitoring as their main strategic approaches. Environmental Defense lauds the impact of "Global Biodiversity Scenarios for the Year 2100," a study co-authored by nineteen scientists from throughout the world. Their findings, published in *Science* magazine, marked the first time an international cast of scientists collaborated in analyzing the human impact on biodiversity loss. As David Wilcove, senior ecologist at Environmental Defense, explains, this type of effort is reflective of what is needed to adequately address the issue of biodiversity protection. Wilcove calls the study "a call to scientists to pursue as much of an interdisciplinary approach to answer the threats of biodiversity loss as possible."[39]

Similarly, World Resources Institute provides technical support to governments, corporations, international organizations, and fellow environmental NGOs.[40] The vast majority of this work, some 90 to 95 percent of it, is conducted in partnership with other organizations. These groups range from large, high-profile international organizations to local conservation or social groups. Through its experiences, WRI found that those who work with other organizations experience a higher degree of effectiveness than those that do not. As Nels Johnson, deputy director of the Biological Resources Program at WRI, asserts, "NGOs cannot operate in isolation from the environment of the communities."[41] The Forest Frontiers Initiative (FFI) provides another example of a partnership at work. In collaboration with IUCN, WRI conducted a five-year, multidisciplinary effort to promote stewardship within biologically vulnerable forests throughout the world. WRI's publication of *Global Biodiversity Strategy* is another example of an effective partnership. This publication was a joint effort of WRI, IUCN, and the United Nations Environment Program (UNEP). Johnson, in fact, points to this partnership publication as one of the more successful efforts that WRI has done in its history. With eighty-five specific proposals for conservation action, the publication sought to stimulate fundamental changes in how local, national, and international levels both perceived and managed resources. The document was released in February 1992 at the World Parks Congress in Caracas, Venezuela, with an eye on the Earth Summit in Rio that summer. It has since been seen as useful on a variety of scales. From multiple applications in the field to the negotiating table itself, *Global Biodiversity Strategy*'s impact can be measured in its virtual omnipresence.

Subsequent translation into thirteen languages certainly shows the demand for local level uses. This is precisely the type of service that WRI offers to the environmental community as a whole. WRI is different than many other environmental NGOs inside the Beltway. They do not have a membership. They have much fewer projects on the ground. And they do not really have satellite offices, with the exception of a couple sporadic staff placed outside D.C. As WRI research analyst Melissa Boness notes, "We are primarily small and here [Washington, D.C.]. WRI occupies a unique place. The tension that exists between other organizations, for all their own reasons, also comes from the fact that they do a lot of overlapping work. And WRI, because of the policy analysis angle of most of what we do, has occupied a niche that is different enough that we have the ability to bring others together without looking like its self-empowering, without looking like we are trying to take over their territory."[42]

The number of joint ventures that WRI engages in over the years demonstrates the validity of this statement. Philosophically, though, as seen in previous discussion of both the domestic-international linkage and the short-term–long-term linkage, partnerships are most influential to the extent that they support those who actually live in an environmentally vulnerable community. WRI does this indirectly when it partners with groups like TNC, CI, and WWF. Of course, those groups do so more directly. The Nature Conservancy's Mary McClellan highlights this point, saying, "If we don't work with local organizations, we are not going to have a long-term impact. All our work is oriented toward site-based protection. We cannot implement a project without working with a local group."[43] The organizational characteristic of maintaining partnerships, along with the aforementioned characteristics, allows NGOs to improve a variety of mainstream strategies toward protecting transnational biodiversity.

In summary, a number of organizational characteristics provide notable support to the mainstream strategies that NGOs utilize in seeking to protect transnational biodiversity. A degree of imaginative decision making, concerted strategic concentration, and the general demographics of staff and membership deserve emphasis. Perhaps the single most influential characteristic, though, is the willingness to establish partnerships with other NGOs. This characteristic provides a remarkable array of benefits. It clearly acts as a valuable support mechanism for NGO strategies.

Supports to Participatory Strategies

Much of the above discussion in the mainstream subsection regarding partnerships applies to the participatory strategies of grassroots networking and community education. Partnerships provide much-needed supports to participatory strategies as well as mainstream strategies. In many cases, this relationship is even more critical, due to the very nature of participatory strategies and the extent to which their effective implementation relies upon communication among a diverse grouping of individuals. This is also true regarding the discussion of the organizational characteristic of general demographics, particularly the measures of membership and staff. Grassroots networking, by and large, is more efficient at membership maintenance than mainstream strategies because members and clientele are already aware of the NGO's actions. The drawback here is that grassroots efforts are inherently unpredictable and at times threaten to undermine the very Washington-based efforts they were originally intended to support. In contrast, mainstream efforts usually go unnoticed without periodic updates by the NGO itself.

On the participatory side, partnerships also provide essential political capital—at times serving as the bridge between the grassroots and more official access channels. Perhaps the best illustration of this is the relationship between World Wide Fund for Nature (WWF-US's international affiliate) and the aforementioned World Conservation Union (IUCN). IUCN is officially an NGO-IGO hybrid with 74 governments, 106 governmental agencies, 35 affiliates, and over 758 NGOs. National and international NGOs thus comprise the vast majority of membership. All total, IUCN has 980 members from 140 different countries. Types of members thus range from environmental NGOs and professional societies to zoos, aquariums, and botanical gardens to academic institutions and indigenous people organizations to governments themselves.[44] Founded in 1948 as the International Union for Protection of Nature (IUPN), it boasts a secretariat staff of some 820 individuals.

WWF is an important member of this unique coalition. In fact, the founding of WWF in 1961 was the brainchild of IUCN itself. Since its inception, IUCN leadership was continually frustrated by the lack of resources to implement its initiatives. WWF was set up to alleviate this need. Its primary purpose was simply to raise money.[45] While WWF has expanded its objectives and responsibilities considerably since its initial formation, it retains a close relationship with IUCN. They share numerous joint program responsibilities, and until the late 1980s, even shared a building at their international headquarters. But the most notable characteristic about WWF today is its global reach. There is a rough division of labor from the two hubs of WWF-International in Switzerland, which acts as the "lead office for Asia and Africa," and WWF-US, which acts as the "lead office in Latin America."[46] However, individual state offices operate under their own accord.

The vice president of the global threats program at WWF-US, Brooks Yeager, notes that WWF today is a true network of forty different national organizations. Each of these stands on its own—but with a number of interlocking programs. Yeager believes this sets WWF apart from other environmental NGOs working in the biodiversity protection arena. He contends, "I think we are unique that way, to establish a global conservation agenda that can be delivered by people in country with local partners . . . Not very many groups attempt to do what we do at the global level. We're the only truly global group that does nature conservation."[47]

WWF is also notable for its partnerships with other environmental-related organizations. WWF and National Geographic, for instance, have formed a joint initiative to combat threats to the world tiger population. Together, these two groups employ community education and grassroots networking as they seek to protect tigers throughout Asia and the Indian subcontinent. Their Tiger 2000 campaign included a Web page devoted to the issue and mailings through WWF's species alert listserv. The expressed intent was to increase the capability of WWF to spread the word about threats to tigers. Similarly, "Weeds on the Web," a collaboration of TNC, the University of California at Davis, and the Native Plant Conservation Initiative (itself a consortium of NGOs and federal agencies) exploited the value of partnerships in linking groups that might otherwise have very little contact with one another. "Weeds on the Web" enhanced the educational strategy being explored at TNC in the late 1990s by providing land stewards with information on how to deal with invasive weeds. The program may also be seen as improving the mainstream approach of scientific and technical research as it directs scientists to collaborative possibilities with others working in their area.

NGOs such as World Wildlife Fund not only use partnerships to enhance education, they also utilize partnerships in improving grassroots networking. Linking up with indigenous groups, WWF broadens its constituency base and improves its ability to make each of the three fundamental linkages. For example, WWF's The Living Planet Campaign targets partners throughout the world in implementing an integrated approach to conservation at the eco-regional scale. As previously discussed in chapter 3, WWF's initiative against illegal poaching and fostering sustainable national forest practices in Zimbabwe is one case in point. In fact, as USAID's Franklin Moore asserts, there is a real shift in this direction, toward using indigenous organizations as the lead implementing entities. As Moore states, "The days of NGOs as direct action agents have been eclipsed by indigenous groups."[48] While many NGOs would take issue with this statement, partnerships with these indigenous groups clearly provide substantial support for the specific strategies NGOs use.

One final example of WWF employing its partnerships is found in its WWF/Clean Energy Agenda program targeting global warming— and the species loss that accompanies it. Several NGOs examined in this study are involved in this coalition, including Environmental Defense and Sierra Club (as well as Union of Concerned Scientists, Natural Resources Defense Council, U.S. Public Interest Research Group, National Wildlife Federation, and Physicians for Social Responsibility). WWF's contribution was enclosing an advocacy postcard as a supplement to *Focus,* the bimonthly publication it sends to all its members. Asking members to mail in postcards advocating clean power, air, cars, and investments, WWF sought to enhance not only mainstream lobbying but also community education and grassroots networking. But more than just WWF's membership was exposed to this initiative. WWF enjoyed the added support from the actions of its partners on this issue. Forming partnerships expanded their constituency base.

The Sierra Club's partnership with Amnesty International USA is another case in point. With each NGO receiving a three-year, $900,000 grant from the Richard and Rhoda Goldman Fund in late 1999, the two can better address their respective core missions by directly targeting the overlap between them. Acting together, the Sierra Club and Amnesty International better protect environmental activists whose human rights are violated. Both Sierra Club and Amnesty International seek to combat human rights abuses perpetrated upon environmental activists. Their partnership is an example of innovative collaboration between two organizations that stress grassroots networking and community education—but in traditionally different issue domains. By coming together with this initiative, titled "Defending the Defenders," both Sierra Club and Amnesty International can improve their abilities to pressure the U.S. government in foreign policy decisions, to target offending governments, and to mobilize grassroots support for activists at risk throughout the world.

As the Sierra Club first stated in its announcement of the program, "In too many countries, it is dangerous business to be an environmentalist."[49] Both groups are using this fund to revamp grassroots and advocacy capabilities. In particular, the grant requires Sierra Club and Amnesty International to increase staffing toward membership mobilization, legislative advocacy, Web site development, and media communications. Most significantly, the grant directly addresses "the gap between funders of environmental issues and funders of human rights."[50] It addresses the ecological-economic linkage and its social base. Partnership with Amnesty International clearly provides valuable support toward strategies utilized by the Sierra Club. Even after this collaborative three-year grant expired at the close of 2002, benefits are still being felt, as Amnesty International now has a full-time environmental program.

The Sierra Club has returned the favor. In December 1999, Sierra Club released *Environmentalists Under Fire: Ten Urgent Cases of Human Rights Abuses*. The documentary film highlights cases in Burma, Cambodia, Chad and Cameroon, China, Ecuador, India, Kenya, Mexico, Nigeria, and Russia. One example is the grassroots effort by the Sierra Club to get its membership to lobby their congressional representatives on behalf of two Mexican environmental activists. In a July 2000 *SF Moderator* message, the Sierra Club asked members to write their representative to pursue the human rights of Rodolfo Montiel and Teodoro Cabrera.[51] These two were imprisoned in May 1999 for attempting to stop forest destruction in Mexico's Southern Sierra Madre. They were then convicted of murder, after a sixteen-month imprisonment and allegedly torture-induced confession, and sentenced to six-year and ten-year jail terms, respectively. Both Montiel and Cabrera were declared Prisoners of Conscience by Amnesty International, who called attention to their alleged torture while being held under concocted charges of illegal possession of weapons and drug trafficking. Over the course of 2000 and 2001, the Sierra Club's Human Rights and the Environment Campaign mobilized substantial grassroots support. These included a rally outside the United States Congress in June 2000, rallies and vigils outside the Mexican Consulates in October 2000, and a rally in Los Angeles in July 2001, as well as thousands of postcards mailed from around the United States to not only U.S. representatives but also to Mexican officials. The two were released in November 2001.

Sierra Club consciously tried to bridge the domestic-international gap in programs such as this. This demonstrates sound political judgment. The environment and human rights are a natural pairing, as noted by William F. Schulz, executive director at Amnesty International USA. Schultz contends, "Those governments which are most oppressive of their human citizens are also most oppressive of their environment, of the entire eco system that is part of their world. The link between human rights and environment is very intimate, very close."[52] Pairing the environment with human rights can also be politically expedient. Getting Americans to care about distant places where the vast majority of species diversity resides is challenging for one reason. It is human instinct to care about what you are closest to.

The environment is a very local issue for most Americans. International program director at the Sierra Club, Steve Mills, explains that Americans care about the water their children drink, the air they breathe, and the beaches where they swim. As Mills contends, international biodiversity "is a tough sell . . . [but you can reach Americans] through the eyes of people doing it, through their stories. Americans may not follow Nigerian political elections, but they do understand when people are under fire. They understand injustice, and that's what this is about." That is where the Sierra Club's Human Rights and the Environment Campaign creates impressive inroads. Mills also recognizes that this adds to the work of other environmental groups without expanding the Sierra Club beyond what it does best, building support from the grassroots up. "We want to help educate Americans. We've got to identify the links through grassroots political muscle. That's where our leverage is because we don't have offices abroad."[53]

The Sierra Student Coalition (SSC) is another example of a unique partnership providing support for strategies employed at the Sierra Club. Founded in 1991, the SSC is the student arm of the Sierra Club. It retains direct affiliation with the Sierra Club but exercises independence in operation. The SSC, with 25,000 members and 250 distinct groups around the United States, is the only national, student-run grassroots network with direct ties to an established environmental NGO. SSC believes this enhances not only their initiatives, but also makes the grassroots strategies that the Sierra Club employs more effective as well. As SSC states on its Web page, "The close connection suggested by our names and structural ties creates a comfortable unity between the Sierra Club and the SSC."[54] The partnership enlarges the Sierra Club's constituency, increasing grassroots networking capabilities and social capital. And it contributes to the long-term strategy of building future Sierra Club membership in younger population cohorts. In some ways, the SSC is a minor league or training ground of sorts where college students such as former Sierra Club president Adam Werbach develop valuable experience.

The Biodiversity Support Program (BSP), begun in 1992, was one final example of a partnership that enhanced the participatory strategy of educational initiatives. Funded by the USAID-led United States–Asia Environmental Partnership and a grant out of USAID's Global Bureau, the BSP included a consortium of WWF, TNC, WCS, and WRI. The program provided funds for some twenty community-based projects in seven countries across Asia and the Pacific before ending in September 1999. Targeting all three fundamental linkages with the strategies of technical support and educational initiatives, the impact of the BSP is still felt today. WWF, for instance, sponsored a BSP Brown Bag lecture series in summer 2000. Held in their D.C. offices, these luncheons relayed research results not only to those in WWF who had an interest in a specific program, but also to other NGO members of BSP and the public more generally. The BSP also made an impact with Biodiversity Conservation Network (BCN), a more locally oriented coalition it spawned. BCN evaluated the relationship between business, the environment, and local communities. Both the strategies of grassroots networking and community education were improved by the partnership nature of this program. Those groups that have an organizational structure conducive to forming and keeping coalitions such as the BSP and BCN improve their participatory strategies effectiveness, and thus improve their ability to meet the three fundamental linkages. Still, there was at least one drawback here, according to CI's Keith Alger. The project never really induced growth or autonomy in local NGOs. Alger believed that BSP was "awfully top heavy [and] spent a lot of its money in D.C."[55]

In summary, both mainstream and participatory strategies benefit from certain organizational characteristics. General demographics of age, financial expenditures, staff, and membership generally enhance the potential of an NGO to implement its strategies when their values increase—with a few notable exceptions. Decision-making style also may act as a strategic support. Those NGOs that are relatively decentralized, transparent, and imaginative enhance their strategic effectiveness. But again, a few caveats must be attached to this blanket statement, particularly as mainstream strategies such as lobbying and litigation require a higher level of centralization than participatory strategies do. The organizational characteristics of targeted constituency and strategic concentration also may support strategies when truly transnational communities are the focal point and the strategies applied fall within the capabilities of an NGO. Finally, partnerships are an integral support mechanism within an organization. They are particularly notable as they assist NGOs in improving both the targeted constituency and strategic concentration characteristics. With this in mind, the following section addresses the conditions under which the above NGO characteristics act as constraints.

STRATEGIC CONSTRAINTS

Constraints to Mainstream Strategies

The demographic of staff, particularly staff stability within an organization, plays a pivotal role in effectiveness. Indeed, high personnel turnover rates create another distinct problem within the mainstream strategies that environmental NGOs utilize. This is most obvious when upper-level managers leave an organization. The Ocean Conservancy faced just such a transition period in the late fall of 1999 as they recruited a new president. Ocean Conservancy's former vice president of marine conservation, Bob Irvin, admits this can be a constraint. He believes that CEO and general staff stability is a definite plus and points to the example of Environmental Defense as an organization with relatively low turnover. According to Irvin, "On balance that (lack of turnover) makes it a stronger organization. If I had my druthers, we would be more like that."[56] In some instances, one organization's loss may be another's gain. And at times the entire environmental community shares a degree of benefit when a staff member moves from an NGO into an access point in a governmental agency. One such example is Brian Day, ironically, a former staff member of Environmental Defense who left that NGO to serve as director of Green-COM. GreenCOM is an environmental education and communication project of USAID. Launched in 1993, it conducts fieldwork and applied research at the request of USAID missions and bureaus. GreenCOM works with government agencies, local community groups, and schools as well as NGOs, although any actions must have host government approval.[57] Still, there is a degree to which personnel turnover hurts individual NGOs in terms of lost networking and lost institutional memory.

This is perhaps most damning in terms of frequency when it comes to mid-level and even entry-level positions. That is, personnel losses occur most frequently among younger, lower-paid staff. When a group is constantly rehiring a slot every two or three years, institutional memory for that particular position obviously suffers. When Earthjustice lost its Washington media consultant in the late 1990s, it took them several years to find a replacement. Over that time period, his files were in their D.C. congressional office, but no one really utilized them because no one really understood them. As Earthjustice's Shana Glickfield, policy analysis assistant within their congressional office, notes, a lot of what she does is about developing relationships, engaging in old-style politics.[58] When people are constantly leaving, the ability to maintain such linkages is jeopardized. This constraint is even more obvious in the recent experiences of Defenders of Wildlife. Defenders lost all of its CITES staff several years ago and is just now getting back into that fora. With this in mind, Conservation International is one group that has made a concerted effort to correct for this deficiency. CI's Cyril Kormos believes that, by treating younger staffers like him as investments, CI serves as a model for other environmental NGOs struggling with this same detrimental phenomenon.[59] Of course, these steps only work as long as advancement is possible within the organization. It cannot, and should not, prevent natural progression up the proverbial career ladder; Kormos himself moved over to the Wild Foundation as vice president for policy in late 2002.

Another constraint of note is lack of strategic concentration, particularly in the mainstream strategies NGOs employ. Too often a group attempts too much, without consideration for the true value-added contribution any new strategies would make. Then again, there are also costs to being cautious in strategy selection. Those organizations that do make a conscious effort to pick their spots carefully run the risk of alienating themselves from the rest of the environmental community. Environmental Defense, for example, is selective in its strategic approach. They pick specific niches with care. But this caution can create frustration in other circles, particularly when it is misinterpreted as a slight against another NGO's favored project. Indeed, Environmental Defense has at times offended other groups by declining to sign on to a particular initiative that another NGO holds dear.[60]

Still, NGOs face far greater constraints when they attempt to extend themselves beyond their original strategic emphasis. Biodiversity Action Network (BIONET) was a prime example. BIONET's original mission was to advocate effective implementation of the Convention on Biological Diversity (CBD) worldwide. Confronted with a United States Senate that refuses to ratify this treaty, BIONET branched out into other areas. Its leadership switched its goal to implementation in specific states and assumed a more open advocacy stance than in the past. Many of its NGO members were frustrated by this turn of events and felt that BIONET had lost its strategic focus. They believe that the monitoring and technical research contributions of BIONET suffered as a result.[61]

The above discussion of strategic concentration is closely related to another organizational characteristic that at times constrains NGO strategic effectiveness. The specific constituencies that an NGO targets also determine its ability to meet its overall objectives. Defenders of Wildlife, like many NGOs, struggles with this decision on what constituency to target. When founded, the primary mission of Defenders was to preserve large predators, specifically to combat the use of steel traps. The original name for the organization, in fact, was Defenders of Fur-bearers, and its logo to this day remains a wolf. Over the years, Defenders' focus broadened to include wildlife habitat and biodiversity more generally. Where Defenders really stretches itself thin, though, is in its attempts to do both domestic and international work.

Despite support of certain high-profile species at the international level, the domestic and international components of Defenders are by no means equal. Much of the NGO's energy and resources is devoted to domestic programs—although member interests force Defenders to address some international issues as well, especially biodiversity protection of high-profile species such as elephants, dolphins, and tropical birds. The international program also devotes attention to international forestry, WTO, and CITES. Given the combination of a small staff and budget, Defenders may be expanding too much in the international arena. As much as Defenders works effectively with other groups, their expanding international program in some ways threatens to undermine these partnership gains. Competition plays an important role here. While Scott Hajost, executive director of the World Conservation Union's U.S. office (IUCN), believes that competition within the environmental community can be healthy, that good ideas can come out of this competition, he also finds too much competition for funding decidedly detrimental. And, indeed, this competition for funding can be intense. The resulting race to write press releases or to line up a certain wealthy donor can create bad will between groups.[62]

As late as 1999, only two staffers at Defenders, Rina Rodriguez and her superior, Bill Snape, were devoted to international issues. And both spent only part of their time in the international arena. In fact, Rodriguez devoted half her time as a Defenders employee in a contracted-out position to Community Nutrition Institute (CNI). Now under Carroll Muffett, Defenders international program has a person on-site in Mexico and plans to hire one on-site in Canada in the near future. And several attorneys such as Bill Snape contribute time to international issues when needed. On the one hand, this expansion is to be applauded, for it rightly exploits the transnational nature of the biodiversity issues that Defenders targets, particularly its focus on wolves since being founded over half a century ago. On the other hand, growing too much could push Defenders into areas better addressed by existing groups. Defenders must also remain cognizant of the resources available internally (as well as the somewhat limited pool of resources available externally). With a staff of around seven plus one administrative assistant, their international program does not have the personpower to incorporate transnational biodiversity protection into the daily agenda, at least not on the order that WWF, TNC, and CI do. Often times, Defenders recognizes this deficiency and relies on partnerships to achieve its objectives as discussed earlier. Yet, perhaps what bothers other environmental NGOs most about Defenders, is the fact that their fundraising rhetoric often gives the impression that high-profile, endangered species found outside the Untied States are a central component of their agenda. These critics believe Defenders gives the false impression that international species are part of Defenders' core mission—and that Defenders' efforts siphon off resources that would otherwise go to them.

One alternative is to take advantage of both an existing strength at Defenders and gaping weakness in the international environmental community at large. No one really addresses the second biggest threat to biodiversity in the world today, invasive or alien species. When non-indigenous species are brought into an ecosystem, either intentionally or unintentionally, the results can be devastating. Defenders has taken the lead on this issue at times, utilizing the several staff members on board with experience in this arena, but too often the issue lays dormant for long stretches of time. Defenders is by no means to blame here. No NGOs are really dealing with this issue, according to a number of government sources. Then again, there is a reason for this. Given the extent of global trade and travel today, many government contacts in the area are unsure whether anybody really can address this issue sufficiently. In many ways, globalization has opened a Pandora's box through the intentional as well as inadvertent transplanting of nonindigenous species, and we are now witnessing the homogenization of the Earth.

Domestically, there is some structure with which to work given the National Invasive Species Council in the United States, which is chaired by the Secretary of the Interior. Set up by executive order by former President Bill Clinton, the body serves largely as just a coordinating entity, trying to get information out to people about the impact of non-indigenous species. What is lacking is real advocacy work that could build on this structure. Perhaps education first, then litigation if needed, would be strategic choices to employ. Some scholars hope that a protocol under the CBD might be developed. In any case, if ever there was a case for education, invasive species are it. It literally takes just one non-indigenous species introduced into a system to wreak havoc. From Asian beetles in Chicago and Latin American fire ants throughout the South to fish that walk from pond to pond in the mid-Atlantic states, the United States alone has its share of problems traced to invasive aliens. Internationally, the stakes are even higher.

Constraints to Participatory Strategies

Constraints to participatory strategies start first and foremost with damaged public image. One measure of this image is a comparison of overhead to funds allocated for conservation. Those NGOs with a low overhead compared to actual money spent on conservation enjoy a better reputation in the environmental community. This reputation acts as a valuable support for the strategies that an NGO selects, particularly when it engages in participatory strategies such as grassroots networking. Examining the transparency dimension of decision making and the financial expenditures measure within general demographics, one can better determine when NGOs actually meet the objectives that grassroots networking sets out to achieve. Some organizations do a better job of allocating their resources to their mission statements than others.

Transparency of these expenditure allocations can enhance a group in the eyes of both its members and the public at large. Activists, quite expectantly, become disillusioned when money originally earmarked for conservation efforts goes instead to fundraising or administration activity. Several measures of note exist. Three used here are GuideStar.com, *Smart-Money* magazine, and the American Institute of Philanthropy Ratings (AIP). GuideStar is a national database providing financial overviews of 850,000 nonprofits. *SmartMoney* magazine evaluates the top one hundred nonprofits in the United States near the close of each year, describing which work efficiently and which spend too much money on administration and fundraising. AIP publishes the "Charity Rating Guide and Watchdog Report" triannually.

As we have seen in chapters 2 and 3, WWF incorporates a variety of strategies, but permeating their organizational culture is a basic faith in grassroots networking. This strategy is enhanced by the fact that for years the great majority of WWF funds have gone into its conservation programs. WWF proudly boasts of the fact that 87 cents of every dollar it receives is channeled into conservation programs. The other 13 cents goes to fundraising (9 cents on the dollar) and management (4 cents on the dollar). Periodic mailings from WWF often include small pamphlets that trumpet these statistics and refer to WWF's recognition by the Wall Street Journal's *SmartMoney* magazine as the top environmental charity in both 1997 and 1998.[63] Unfortunately, WWF has lost this title over the last several years and slipped to the fifth slot of the six environmental charities that make *SmartMoney's* list.[64] WWF only placed sixty-nine out of one hundred charities in 2002.[65] TNC faired even worse, placing number eighty-four and dead last among the six environmental groups listed. In fairness to this low score, though, at least according to TNC, their land purchases are counted by the Internal Revenue Service as administrative rather than program expenses, and *SmartMoney* uses the IRS figures.

More exhaustively, table 13 reports data from AIP on nine of the eleven environmental NGOs this study examines. Only BIONET, which is now defunct, and Earthwatch are omitted.[66] The nine NGOs studied that are included exhibit a range in the four AIP categories. The first of these addresses the open-book policy of each NGO. It indicates whether documents were provided to AIP and may serve as one measure of transparency. Documents here include the annual report, complete audited financial statement, and Internal Revenue Service form 990 with Schedule A when applicable. Second, AIP measures the percentage of funds collected spent on the intended charitable purpose. A target of at least 60 percent is recommended by AIP. Third, statistics are given on fundraising costs. AIP be-

Table 13. AIP Ratings for Environmental NGOs

	Open book	% spent on charitable purpose	Cost to raise $100	Grade
CI	Yes	89	$5	A
Defenders	No	42–73	$20–$50	D
Earthjustice	Yes	72–73	$14–$15	B+
Environmental Defense	Yes	76–81	$14–$18	A–
TNC	Yes	79	$10	A–
Ocean Conservancy	Yes	65–72	$20–$26	B–
Sierra	No	62	$33	C+
WRI	Yes	78	$6	A–
WWF	Yes	70–81	$21–$38	C

Source: American Institute of Philanthropy, "Charity Rating Guide and Watchdog Report," no. 32, November 2002.

lieves that the cost to raise $100 should be $35 or less. One should also note that the AIP figure here is often different from that which environmental NGOs themselves may offer because AIP compares fundraising expense to related contributions. That is, AIP considers money received as a result of fundraising activities whereas some groups often compare their cost to raise money with total income, including investment income, sales proceeds, and the like. This can make an organization appear more efficient with its fundraising than it truly is. Finally, the first three statistics from above determine the fourth and final column, the actual numerical grade that an NGO receives. Before totaling these, though, AIP also penalizes groups for having excessive reserves. Those groups that hold assets equal to three to five years of operating expenses receive a grade reduction. None of the environmental NGOs above were in this category in late 2002.

Table 13 shows that Conservation International with an A and Environmental Defense, The Nature Conservancy, and World Resources Institute with "A−s" all recorded exceptional marks. Each had at least 76 percent of their funds spent on conservation objectives. And each spent 18 percent or less of their member donations on fundraising activities, with CI leading the way by spending only $5 for every $100 raised. Defenders rests at the other end of the spectrum, receiving the worst marks and the only unsatisfactory rating. That NGO shows closed books, a wide range reported on the percent of funds that actually went to conservation programs, between 42 and 73 percent, and a relatively high range for the amount of money required to raise $100 on average, between $20 to $50. All this factored into the D they received for their overall grade.

In fairness to Defenders, the difficulty in distinguishing between fundraising and educational initiatives may explain part of this low grade. In several cases, for instance, a range of numbers exists for the category explaining the percentage spent on charitable purpose. Direct mail and telemarketing solicitations are classified as fundraising costs in the lower number for each range. Admittedly, mailings and phone calls may serve a dual purpose. They both raise funds *and* educate various constituencies. The higher number in the range for this category reflects mailing and phone calls that groups claim as primarily program expenses, the cost of acquiring new donors or members as a program service. Thus, those groups that emphasize community education as a strategy exhibit a much wider variance. Still, Defenders range is abnormally high when compared to the other NGOs. Furthermore, the low end of the range on "% spent on charitable purposes" is dismally low at 42 percent, and Defenders was one of only two groups here to close their books.

In the middle of the pack, WWF and the Sierra Club record relatively average scores, at C and C+ respectively. Several groups have improved in AIP's standing over the last few years, including Earthjustice, Environmental Defense, Ocean Conservancy, and the Sierra Club. WWF, on the other hand, has slipped. Mirroring their drop in *SmartMoney* magazine, WWF fell from one of the top AIP-rated environmental groups with a B+ in Spring 2000 to an average score of C by the end of 2002. Increased fundraising costs are likely the prime culprit here, as WWF at times uses as much as $38 to raise $100. The Sierra Club, while improving, is only slightly above average for similar reasons. It uses just under the recommended level of funds to raise money, spending some $33 for every $100. The Sierra Club also failed to open its books to AIP in 2002.

The above numbers reflect the extent to which some transnational environmental NGOs struggle with organizational maintenance. NGOs must be cautious not to let their age mutate from strength to weakness. On a positive note, age clearly enhances adaptability, as political scientist Samuel Huntington has asserted in his conceptualization of institutionalization.[67] Organizations that grow over time develop remarkable benefits in terms of prestige and reputation.[68] Still, there are notable dangers that an NGO encounters as it ages, namely the functional age that Huntington identifies. When an NGO takes on a life of its own, its original mission may be subsumed by organizational maintenance. As NGOs expand in terms of both

agenda and staff, for example, they often assume some of the very characteristics that handicap states in their pursuit of effective transnational biodiversity protection. They become more bureaucratic. They become slower.
They become more state-like. Domestic interest group literature points
out the pitfalls that often arise here as NGOs mature. Particularly in the
case of large, transnational organizations, NGO executives must be mindful of the extent to which the demands of organizational maintenance encroach on their ultimate goal, effective biodiversity protection. Spending
significant portions of organizational funds simply to raise additional funds
and cover administrative costs obviously limits the amount of money available for actual conservation initiatives.

Former USAID democracy fellow and current professor of government
at American University Ronald Shaiko finds that environmental groups in
the twenty-first century are faced with a much different organizational climate than that found in the 1960s.[69] Shaiko identifies the "demands of organizational maintenance" as an increasingly time-consuming element of
environmental NGO daily business. Maintenance has become so critical in
NGO operations that it threatens to encroach upon the actual initiatives
that NGOs sponsor. As they evolve, changing from smaller, grassroots orientations to larger, more professional and mainstream approaches, NGOs
confront a difficult choice.[70] Too often, groups cannot both maintain
growth in their organizational structure and provide effective political representation in the policy-making process. They must choose between one
or the other.

On the one hand, NGOs that begin to resemble the state agencies and
departments they seek to influence do gain certain advantages in pursuing
mainstream strategies. By "scaling up" and becoming more professional,
NGOs are able to enter discussions on a somewhat more equal footing.
They are able to address the root international causes of local problems.[71]
On the other hand, NGOs must also recognize the constraints this growth
creates, particularly in participatory strategies. Time and resources allotted
to simple organizational maintenance may become an undue burden,
threatening the ability of an NGO to devote the requisite amount of resources to its original mission. When NGOs fail to compensate for this deficiency, furthermore, they inevitably acquire the same parochial interests
that limit individual states. NGOs must constantly guard against this tendency if they are to remain truly independent sources of power.

Perhaps the most notable limitation that participatory strategies face is
that of targeted constituency. While NGOs using mainstream strategies
struggle with this constraint, as demonstrated earlier, those using partici-

patory strategies struggle even more. This may be explained simply because
the initial emphasis here is on people actively involved in civil society. Mainstream strategies, on the other hand, stick to official state channels. Regarding the one global biodiversity treaty in existence today, the CBD, the
NGO community has "dropped the ball" when it comes to constituency
development—let alone maintenance.[72] Even the Sierra Club admits to
this constraint within their organizational structure. No viable domestic
constituency for transnational biodiversity protection exists within the
United States. The primary reason for this is simply that no one benefits
politically—in the short term. At the collective level, biodiversity protection benefits apply to all. But given the complex set of interrelated relationships that surround this issue, individual benefits are often only realized over long periods of time. No politician will expend the political
capital needed to push through an agreement in this climate, because our
political process fails to extend its rewards beyond the short term.[73] To
shepherd the level of change needed, NGOs must provide incentives for
politicians to initiate policies with a longer time horizon. They must help
mobilize a viable constituency that rewards forward-thinking politicians.
To do this, NGOs must better incorporate constituency targets and emphasize the direct stake that individuals within the United States have in
biodiversity protection halfway around the world.

This is no easy task. Promoting an understanding, let alone motivating
action toward outright protection of species diversity, is perhaps most difficult in the United States. A vast majority of the United States population
is not directly dependent upon farming or animal husbandry and management on a daily basis. People do not comprehend the relationships involved here. They need to learn about the process, to interact in a manner
that allows for perspectives to evolve. This is where participatory strategies
have enormous potential, but only when organizational characteristics
such as targeted constituency are further adjusted to foster the educational
initiatives needed to increase understanding of the relationships involved
in this complex issue.

Finally, strategic concentration is of note. Wandering from strategy to
strategy creates unnecessary constraints upon the ultimate objective of transnational biodiversity protection. Community educational initiatives require
a dedicated persistence to be effective. Those that are assembled ad hoc
lack the necessary force to get their message across. Similarly, those that are
delivered inconsistently become less of a staple to the information diet that
a community consumes in its daily or weekly decision making. This can
have an impact even when an NGO is targeting constituencies that are

physically distant from the actual conservation sites. For example, *Nature News,* the monthly electronic newsletter of The Nature Conservancy, was postponed in the late spring of 2000 as TNC spent their energies on revamping their Web site. This interruption in service not only forced activists to search elsewhere for information that they needed, it damaged the grassroots networking TNC has only recently begun to apply in shoring up its core strategies of research and property acquisition and maintenance. TNC would probably be best served by either steering away from such a participatory strategy altogether or sticking to regular delivery of their activity updates.[74] The lack of strategic concentration constitutes a constraint.

SUMMARY AND FINDINGS

In summary, this analysis shows that, if environmental NGOs wish to maximize their effectiveness at transnational biodiversity protection, they must first recognize that several basic organizational characteristics support the specific strategies they employ—and that several other basic organizational characteristics constrain them. The main thrust of this chapter is that general demographics, decision-making style, partnerships, targeted constituency, and strategic concentration all have an impact upon effectiveness. Partnerships, in particular, play a pivotal role. The majority of environmental NGOs cannot be effective on their own. Lacking organizational resources at one time or another, they must forage for help. They must network. They must form coalitions. They must rely on partnerships with other NGOs. Armed with this characteristic, the willingness to form partnerships, NGOs can better adjust other characteristics such as strategic concentration, so that they too enhance effectiveness. Environmental NGOs realize that, just as no group wins all the public policy fights it enters, no one strategy wins all the time either. Multiple strategies are thus employed to increase the probability of being effective. Yet employing too many strategies within one NGO can be counterproductive. Despite the best of intentions, some NGOs are now bloated with more strategies than they can handle.

Perhaps the most damning obstacle that NGOs face, then, is one of their own making. Initial short-term successes in terms of absolute organizational growth carry a significant amount of baggage, namely as seen in the pressures to further perpetuate existing programs and even to expand into new strategic arenas. This is where an NGO must retain a constant vigilance regarding its basic demographics. NGOs must remember that one important characteristic distinguishing them from states is their speed and agility. Initial successes may encourage first growth, then further fragmentation of institutional energy. This siphons off valuable resources for simple organizational maintenance, resources that would otherwise be available for specific mission objectives. Paradoxically, NGOs run the risk of becoming victims of their own successes. In struggling with their initial growth spurt, NGOs must be mindful of these pressures and not lose sight of their original issue focus.

This task cannot be understated. Demands on the time, money, and energy of an NGO will come from both outside and inside the organization. As illustrated in the case of Defenders of Wildlife, periodic surveys of their membership often pinpoint high-profile species such as elephants, tigers, and tropical birds as worthy conservation targets. Defenders is not properly equipped to engage in the massive transnational conservation efforts that the protection of these species requires. Defenders' efforts would be better spent, instead, focusing upon the domestic initiatives in the United States for which they were originally founded—or as noted earlier, in the underdeveloped international arena of invasive species threats to biodiversity. Similarly, staff within an organization may push for specific strategic initiatives that extend an NGO too far from its original objectives. At times, personal interests dictate a change in emphasis, but at other times programs are initiated merely to keep up with other groups in the environmental movement more generally. NGOs that lack safeguards against these pressures to keep up with the Joneses often damage the vitality of their organization overall by trying to assume too many responsibilities. They try to tackle too many strategies, and quality suffers as a direct result. While this is by no means intentional, it can be devastating nonetheless.

Fortunately, NGOs have begun to realize the dangers of spreading themselves too thin, of trying to take on too many issues and too many programs. Strategic focus and value-added initiatives are the new buzzwords in the community—and rightly so. NGOs must take a long, hard look at both their organizational objectives and their organizational structure. They must assess the extent to which these two items support each other and adjust accordingly. As this chapter suggests, despite their best intentions, organizational characteristics often thwart organizational objectives. Recognizing this threat and incorporating more of the support mechanisms outlined above, especially partnerships and people-oriented programs, makes the likelihood for effectiveness all the greater. Effectiveness of NGOs is intrinsically tied to this function. Continual self-assessment combines with a well-honed awareness of other transnational environmental actors, specifically other NGOs, to enhance the effectiveness of a given NGO. It allows NGOs to fill existing vacuums in the movement. It increases both their potential to perform much-needed tasks on a case-by-case basis and their ability to provide long-term support to communities and other NGOs in specific strategic arenas.

Granted, there are benefits to self-sufficiency. Still, much redundancy can be eliminated in the transnational environmental community of today. Too many times, departments within one group insist on "re-inventing the wheel" when that valuable time, energy, funding, and, perhaps most importantly, social capital could be better spent in another arena. Presenting a more united front also makes a stronger impression to constituent communities, both domestic and international. It communicates more directly the message that transnational biodiversity transcends particularistic interests. It allows environmental NGOs to build an organized whole greater than the sum of its individual parts. As NGOs wear different hats, over time, even the most disparate NGOs have something to offer each other. It is in this context that this chapter combines examination of the mainstream and participatory strategies with organizational characteristics.

NGOs must find their own strategic niche within the context of what other groups are offering. Yes, at times it is helpful to add amplitude, to increase the crescendo of voices, by copying specific programs and initiatives. In very basic terms, advocacy often means more when more people push it. Yet, there is a fine line between substantive contributions and mere noise. NGOs must resist calls to join a chorus of "me-too's" in the environmental community. Truly effective NGOs carve out their own strategic niche. They consciously work on bettering themselves. This is where the basic organizational characteristics of an NGO act as a filter. Those groups that are able to utilize partnerships, target a transnational constituency or even one specific domestic constituency, and frame their strategic agenda within a value-added context will find this niche. They will be more effective.

Epilogue:
Building the next ark

A CHANGING WORLD

Technological advances, namely the globalization of communication through such dynamic mediums as the Internet, have allowed NGOs to mobilize broader constituencies than ever before—and at times deeper constituencies as well.[1] By exploiting these new mediums, environmental NGOs have enjoyed clear benefits from globalization. Epistemic communities transcend national borders now more than ever. Of course, these advances are tempered by notable costs. The rise of the transnational corporation and subsequent globalization of the international political economy in the 1980s and 1990s exacerbated many problems that these NGOs sought to solve—and launched such new planetary threats as global warming and the issue of species loss, which this study targets. While environmental NGOs did not become weaker in absolute terms, compared to the 1970s, environmental groups did lose strength in relative terms. The heightened mobility of capital facilitated global trade and investment. It also allowed corporations to take advantage of the enhanced leverage at their disposal vis-à-vis governments, both domestically and internationally. Environmental NGOs reacted slowly to this development. Much like other groups, environmental NGOs were caught off guard by the challenges that globalization presented. After a series of victories such as the Clean Air Act and Clean Water Act in the United States during the 1970s, environmental groups probably became complacent as well, thinking they had already established themselves as a significant force on the battlefield—and not realizing that a new battlefield had been created.

As a new generation of ecological concerns arose within this globalization framework, environmental NGOs scrambled to adjust. Like many other organizations, environmental NGOs struggled with the overarching theoretical obstacles to a set of issues that transcend international borders.

Realist state sovereignty concerns clashed directly with liberal cooperation needs as actions or events in one state often impact other states halfway around the world. Being physically distant in one region no longer made one immune to the consequences of events in another region. In September 2000, for instance, researchers found that coral-damaging fungi in the Caribbean Sea could be traced to dust blown from parched lands thousands of miles away in Africa. Led by the U.S. Geological Survey Center for Coastal Geology, this team of researchers found that long droughts in Africa not only had disastrous impacts on that continent. Several hundred million tons of eroded soil annually spread to the Western Hemisphere and damaged coral reefs there.[2]

This interdependence and vulnerability was made even more clear the following fall. In September of 2001, as we all well know, tsunami-size shock waves were sent throughout the globe when al Qaeda terrorist attacks on the World Trade Center in New York City and the Pentagon in Washington, D.C., killed more than three thousand people. While international terrorism had been brewing for years—even decades—before that fateful date of September 11, 2001, this was the first fatal attack on American soil. The attack made at least one point painfully clear. Americans finally realized as a society that the ambiguity of terrorism in no way rendered the concept meaningless, and that, as we entered the twenty-first century, the world was truly a much smaller place—one where global connections make everyone vulnerable.[3] The controversy that swirls around terrorism, including both immediate and long-term responses to the September 11 attacks, have had and will continue to have a profound impact on democratic societies the world over. As many a newscaster has commented, the world will never be the same.

This can be for the better or for the worse, both in terms of the battle against terrorism and the "fight" for biodiversity protection. Attempts to end habitat destruction, invasive species intrusions, and general pollution could benefit from appropriating various models of the war on terrorism. These must be multilateral models, however, not unilateral endeavors. Then again, there is also a danger that preoccupation with this imminent threat of international terrorism will push the "fight" for biodiversity protection even further into the far recesses of our collective consciousness. Let us hope that this latter scenario is not the case, for, if it is, the possibility of an even more terrifying future is very real. Potential ecological irreversibility, in fact, threatens to render insignificant any debate as to what the appropriate response should be. Waiting much longer to coordinate a response could well mean any action at all comes too late.

Here is where historical perspective is useful. There are examples of monolithic threats to the vital interests of the United States from the past. And there are examples where virtually every sector of American society organized around a common effort to prevail over that threat. That is what is needed today. Entire societies must commit themselves intellectually and financially to the effort. Yet this is rarely done. Only in extreme crisis events can we point to where just one society, let along a number of different societies, organized with a common vision to achieve a greater purpose. The most powerful precedents are found in the world wars of the twentieth century, particularly World War II. This is the example to which former Vice President Al Gore points in his call for a global environmental Marshall Plan, one where all of American society would be organized to combat international environmental threats such as global warming, biodiversity loss, and worldwide ecological degradation. War was indeed a tremendously unifying force in the previous century, as evidenced by the Soviet-American alliance against Nazi Germany.

And, despite deep divisions internationally before and after the United States invasion of Iraq and removal of Saddam Hussein in the spring of 2003, perhaps the twenty-first century will still be one where states join hands in a war against terrorism. But in the case of species loss, the challenge is even greater. This issue is complicated by the fact that the enemy is not a dastardly, amoral face akin to Adolph Hitler or the resolute hate within Osama bin Laden. The enemy is within. It is you and I. We are the enemy. We are the root source of this planetary plague. We are the ones responsible for the habitat loss, invasive species introduction, and general pollution. But it does not have to be this way. We can be part of the solution instead of the problem. In fact, ultimately, the only solution depends on us. Community by community, people must move from opposition to support.

WHY NGOS

Nongovernmental organizations provide the medium to achieve this. Transnational biodiversity protection is both complex and interdependent. Many issues are at stake and many actors are involved. But one set of actors in this story, nongovernmental organizations, stands out due to their unique ability to foster civil society by operating both above and below state governments. NGOs thus shape biodiversity protection efforts better than any state or group of states could on its own. This is all the more relevant today, when NGO presence is found at nearly every stage of the policy-making process. Groups shape policy from genesis to implementation. They even contest decisions after policies are enacted into law. And again, NGOs are unique in that all this activity occurs both inside and outside the state. Yet, while they seek to influence states, NGOs are, by definition, nongovernmental. They remain distinct from the governments they seek to influence, even when they serve on state delegations at various international conventions or official conferences of parties.

But before delving further into these responsibilities, one more important backdrop must be noted here, namely the collapse of the Soviet Union in 1991. An entire cognitive framework built around the constructs of the Cold War collapsed along with the Soviet Union. This is indeed notable for transnational biodiversity protection, because the fall of the Soviet Union allowed previously suppressed issues to enter center stage in the international arena. The sudden conceptual vacuum allowed nonmilitary, security issues to receive greater attention. This attention, in turn, exposed the dangerous cognitive dissonance within sustainable development. Thinking strategically, environmental NGOs may further exploit the void left with the end of the Cold War. Renewed attention to terrorism might hinder attempts here, but it should not. As the Sierra Club's Steve Mills asserts, "We know that environment and security are related. [We are discovering that] destroying the environment abroad impacts security back home."[4] Uncontrolled development and its relationship with human consumption, in fact, sow many of the seeds of hate for the West today. Perhaps the threat of transnational biodiversity loss (as well as global warming, ozone depletion, and other global environmental threats) is not as obvious or as immediate a threat. But it is a security threat, one that could eventually make international terrorism pale in comparison.

To avoid this nightmare, the environmental community must learn to tell a better story about the crisis of transnational biodiversity loss. If citizens and their respective governments are going to have any chance of protecting this precious, global resource, an even more convincing explanation of both the benefits this resource provides and the threats that confront it is needed. Yet to reach this final objective, NGOs must first tell a better story about themselves. Telling a better story about global environmental issues such as biodiversity protection entails telling a better story about the ways that NGOs contribute to governing state institutions, international agreements, and civil society more generally. Storytellers must first establish the proper setting and introduce themselves, warts and all. Meticulous attention to detail in these early stages makes possible full realization of the potential sway a story holds for the long term. This is precisely the case regarding the issue of biodiversity protection. Only by examining the strategic initiatives that NGOs employ can we understand the most effective route to protecting transnational biodiversity. Such an analysis allows NGOs to become better storytellers, to continue developing their status as the conveyors of knowledge and expertise.

This is true worldwide, including within the United States. With vast technical and scientific expertise at its disposal, the United States turns to NGOs for these services less than smaller, poorer states do. The "clienteleism" academic Theodore Lowi asserts that domestic interest groups enjoy has yet to be found within the NGO-state relationship.[5] Yet NGOs do enjoy certain privileges in various international settings and serve as valuable support networks for both small and large states, the United States included. They utilize a variety of access points, particularly in the United States, as seen in the discussion of the mainstream strategy of lobbying in chapter 2, even at times winning support within an agency. This support, however, does not constitute the type of illicit relationship that Lowi postulates. The ecological-economic linkage that environmental NGOs pursue helps explain this distinction. While interest groups often represent strong economic interests for a specific constituency, NGOs do not. Yes, they can represent economic interests, but these interests are intermeshed with other interests such as ecological or social interests—and the economic interests that NGOs represent often transcend national borders. They target global public goods instead of particularistic goods. This analysis, then, also stands in stark contrast to the conclusions of renowned economist Mancur Olson. Whereas Olson believed that political freedom allowed groups to produce economic stagnation that would ultimately threaten democracy, this book concludes that a democratic culture of political freedom allows environmental NGOs to combine mainstream approaches with participatory approaches so that ecological (and economic) stagnation does not threaten democracy.[6]

NGOs thus are a crucial component of the design team in constructing the next ark, even acting as the lead carpenter at times. As political scientist Russell Dalton proclaimed in the late 1980s, environmental groups have become pivotal new players in the policy process.[7] Dalton believes citizen politics is a key element in the political process conducted within contemporary democracies, that how individuals view politics and participate in the process is the definitive variable in determining effectiveness of a policy issue.[8] This book takes a similar line of argument. It contends that environmental NGOs are an increasingly popular choice of medium for the citizen politics that Dalton addressed—and that NGOs are a key determinant in the effectiveness of transnational biodiversity protection efforts. They are a fundamental component of the political process in contemporary democracies and transnational relations more generally. How NGOs view politics and participate in the process is important.

Former UN Secretary-General Boutros Boutros-Ghali agrees with this assessment. He envisions NGOs as critical agents of mobilization in a number of international activities, including confronting environmental problems such as transnational biodiversity protection. This is an especially vital function, he believes, because NGOs serve as the day-to-day link between democracy and peace.[9] As Boutros-Ghali argues:

> Nongovernmental organizations are a basic form of popular participation and representation in the present-day world. Their participation in international organizations is, in a way, proof of this. It is therefore not surprising that in a short time we have witnessed the emergence of so many new NGOs, which continue to increase in number on every continent.[10]

NGOs do not only represent democratic participation at work, though. They also make democratic participation possible. NGOs allow those who would traditionally be excluded from the decision-making process to have a say. As Boutros-Ghali explains, NGOs generate the political will needed to make tough decisions because they are seen as true representatives of the stakeholders on an issue. As such, they wield substantial mobilizing powers.

FOSTERING MORE PARTICIPATORY DEMOCRACY

This is precisely where the theoretical lens of participatory democracy holds enormous power. Participatory democracy provides a rainbow of hope.[11] It provides the political context for building the next ark, for weathering the biblical deluge that, once again, man has brought upon himself.[12] As noted in Genesis, Noah built his ark to withstand forty days and forty nights of rainfall, a storm that was God's punishment for man's wickedness. As we enter the twenty-first century, human production and consumption levels continue to rise at a level that chokes out many other species. These practices raise the fear that our species will be punished once again for its "wickedness." This time around, moreover, humankind appears to be arrogantly and ignorantly playing the role of God as well. We are the ones who are unleashing the torrential downpour, the flood that requires a new ark.

Environmental NGOs help build this new ark by continuing to develop their ability to convey knowledge and expertise. They perform this task using a variety of strategies, both mainstream and participatory. Yet, NGOs can only be effective if they employ these strategies within the context of participatory democracy. Looking to the discussion of ecological literacy by Oberlin College's David Orr provides some insight here.[13] His emphasis on a three-tiered approach incorporating knowing, caring, and practical competence about environmental relationships explicitly emphasizes experiential education. It means those closest to the environment have just as much, if not more, to share in terms of knowledge with those that bring technical expertise and financial capital to the table. It means participating democratically in the decision-making process is key. It means both Northern- and Southern-based NGOs have insight to offer one another. It means that addressing the domestic-international linkages is critical to protecting transnational biodiversity.

But Orr's discussion of ecological literacy and the participatory environment in which it develops also entails connecting the ecological and economic considerations that permeate our issue. And it requires connecting the short-term and long-term considerations as well. This, too, is where NGOs provide a valuable service. Without many of the parochial constraints of other international entities, NGOs can negotiate more neutrally the trade-offs that, at times, truly do exist between conservation and economic development. As Peter Herkenvath of Birdlife International explains, "The CBD needs NGOs more than any other convention. They improve both input and output. They extend expert support beyond science to include social science."[14] Similarly, as USAID's Franklin Moore notes, "Our best access comes when environmental and developmental groups ally." While Moore represents a United States government-lending agency, USAID works closely with a variety of NGOs, using them as a conduit to bypass the corruption and red tape endemic to many third world governments.[15] By funneling funds through its New Partnerships Initiative, USAID strengthens NGOs, which, in turn, strengthen grassroots democracy. As Moore explains, this has translated into some notable successes. "[Programs on] the environment [have] done as much for democracy and governance as democracy and governance [programs have] done for democracy and governance," he asserts.

In incorporating the rhetoric of democracy, though, one should be cautious of the fact that NGOs are "not . . . elected or necessarily accountable to the peoples whom they claim to represent."[16] This word of caution cannot be taken too lightly. NGOs are under constant pressures to streamline their decision-making procedures so that they may more efficiently respond to governmental actions. Cutting participatory elements allows NGOs to act quicker—but at exorbitant costs. Shortening the decision-making process eliminates the very channels of participation that NGOs were created to provide. And within a wide array of issues, these actions of political expediency ultimately reduce the original political purposes of the organization. Indeed, NGOs cannot truly establish any of the three linkages needed to protect transnational biodiversity effectively without some degree of participatory democracy guiding their efforts.

BUILDING THE NEXT ARK

No one strategy, no one style, is the "right" way to protect transnational biodiversity. As seen in this analysis of eleven different environmental NGOs, many options for contributing exist. The comparative cases of this book demonstrate that the greatest strength of the environmental movement rests in its diversity. As BIONET's Hans Verolme asserts:

> We have been assigned a particular task for the environmental movement as a whole. The little bit of work that I do fits into a much, much larger picture—even though it might not always be as coherent or forward-looking as I may like.[17]

One can wear suits or don camouflage, as former Earth First! member Dave Foreman has said.[18] Both uniforms have influence and contribute to the process. What this analysis does determine is how different strategies fit within the overall objectives of an NGO, how effective strategies are in achieving the three fundamental linkages, and what role particular organizational characteristics play here.

Environmental NGOs have different goals, philosophies, styles, structures, and methods. At times this can be a decided weakness, as fragmentation of the environmental NGO community can stymie presentation of a united front to policymakers.[19] Still, this book finds that, at least in the realm of strategies and tactics, diversity within the environmental NGO community is welcome. Indeed, the preceding pages argue that the participatory strategies of grassroots networking and community education enjoy a symbiotic relationship with the mainstream strategies of lobbying, litigation, scientific/technical research, property acquisition/maintenance, and monitoring. Together these strategies increase the ability of NGOs to make the three stated linkages of domestic and international, ecological and economic, and short-term and long-term interests. That is, participatory strategies add to the more traditional mainstream approaches by involving the requisite civil society whereas mainstream strategies add to participatory approaches by directly accessing the state.

In conclusion, an effective biodiversity protection program must support both the environment *and* the people by targeting both the people and their governments. Whether this positions Stephen Kellert and Edward Wilson's political idealism of biophilia or a more realist, long-term economic perspective as its foundation is immaterial.[20] Quite simply, as Peter Raven, the Missouri Botanical Garden's director, asserts, "There will be no adequate preservation of biodiversity without the attainment of ecological stability—sustainability—throughout the world."[21] Information is the key here. NGOs are the conduits providing this much-needed enlightening function. They connect the nation-state to grassroots movements and to the global arena. As Jeff McNeely, founder of the Global Biodiversity Forum and chief scientist at the World Conservation Union, convincingly argues along with several IUCN colleagues, the more widespread environmental knowledge becomes, the more likely solutions will be possible. This is the best alternative for those seeking to strengthen transnational biodiversity protection and international policy agreements such as the CBD. To be effective, biodiversity management must apply information-spreading techniques, albeit in consultation with the existing conceptions of sovereignty that complex interdependence allows.

It is within this context of the need for a broadened political spectrum that this analysis also contributes to participatory democracy. Admittedly, a host of obstacles confront contemporary environmental public policymakers. The most noteworthy of these are scientific uncertainty, relative or even unknown value, widely divergent scope and scale, fluctuating timeframes, and basic deficiencies in public attentiveness. Yet, concerted attempts to open up the decision-making process more fully to the general public can often overcome these hurdles. Utilizing the benefits of a truly democratic political discourse, in fact, is the best alternative here. Effective communication must go beyond the traditional powers format and incorporate alternative, more diverse, and deeper sources of knowledge and knowledge dispersion made possible largely due to the Internet revolution.[22] As John Dryzek, head of Social and Political Theory Program at Australian National University, outlines, communication applied in this discursive manner is more environmentally benign and brings previously unheard voices to the table.[23] It makes implementation of sustainable environmental policies possible. It makes truly transnational biodiversity protection feasible.

NGOs must continue to incorporate this theoretical lens in their day-to-day operations and their long-term agendas. This is no easy task. And the stakes are undeniably high. A new storm is brewing as dark clouds amass at the aforementioned estimated rate of fifty thousand species extinctions per year. We need a contingency plan. There will be many actors in this effort. State governments and corporations are two such obvious categories. But there is often a need for a different source of power, a nonprofit source of power that can operate both above and below the state. NGOs provide this outlet. And they have demonstrated the ability to both work with corporations and challenge them at the same time. Environmental NGOs are instrumental in the effort to construct the next ark. But to be effective, they must carve out strategic niches. They must work together, resisting the temptation to treat a limited pool of funding and public exposure as a zero-sum game. And they must more consciously incorporate the three fundamental linkages of domestic-international, economic-ecological, and short-long term interests outlined in this analysis. Only then will NGOs truly help build the next ark.

Appendix A
NGO case selection

Organizations were selected for this study based on variation of their strategies, diversity of their organizational characteristics, and reputation for past effectiveness. Selection covers Porter and Brown's three main divisions of environmental NGOs (international, large national, and think tanks) as well as an array of ideological orientations. The eleven groups, while all holding both a national presence in the United States and international interests specifically in biodiversity protection, are remarkably diverse. Starting with general demographics, the groups range in age from the Sierra Club, which was founded in 1892, to Earthwatch Institute and the Ocean Conservancy, founded in 1971, to the short-lived Biodiversity Action Network (BIONET), operating from 1992 to 2000. Also of note is the fact that almost half (five) of the NGOs studied here were founded in the 1960s and 1970s (with another three founded in the 1980s and 1990s), a concentration reflective of the environmental movement at large. Other demographic measures, namely expenditures, staff size, and membership size, also vary considerably. NGOs such as Environmental Defense, Earthwatch Institute, and the Ocean Conservancy represent the smaller end of the spectrum on each of these measures, while World Wildlife Fund, the Sierra Club, and The Nature Conservancy represent the larger end.

Similarly, decision-making style varies widely across two of the three dimensions examined. NGOs such as Environmental Defense and Defenders of Wildlife exhibit high centralization, whereas the Sierra Club and The Nature Conservancy exhibit low centralization. Most groups did demonstrate high transparency, although Defenders of Wildlife demonstrated less than others, as seen in their refusal to provide the American Institute of Philanthropy with various financial materials (see table 13 in chapter 4 for further description here). Imagination varied. Grassroots groups like the Sierra Club and World Wildlife Fund as well as more mainstream strategy–oriented groups like Environmental Defense scored high. On the other hand, NGOs formed for a specific strategic function scored low on imagination. But as argued in the strategic concentration section, this is not necessarily a constraint.

The third organizational characteristic identified, partnerships, displays the least variance of those studied, although NGOs do vary to the extent that they emphasize networking and developing coalitions as part of their core mission. For instance, WWF, CI, Environmental Defense, and Sierra Club all have displayed a unique ability to exploit this characteristic. As symbolic of NGOs more generally, though, all environmental NGOs in this analysis attempt to integrate the supports that partnerships bring to their specific strategic initiatives.

The same cannot be said for targeted constituency, the fourth and perhaps most challenging organizational characteristic selected in this study. NGOs such as Defenders of Wildlife almost entirely target domestic constituencies in the United States. The Sierra Club also emphasizes domestic issues in the United States, but increasingly incorporates international issues in an attempt to stake out an international constituency. Environmental Defense and the Ocean Conservancy similarly struggle with the domestic-international tradeoffs, in the end leaning more toward an international agenda. At the other end of the spectrum, WWF and CI target a combination of domestic constituencies throughout the world in an attempt to mobilize a truly transnational constituency.

Finally, the fifth organizational characteristic, labeled strategic concentration, also varies, both in terms of what strategies are actually emphasized and precisely how many types of strategies are emphasized. Strategies range from mainstream to participatory and unique combinations of the two. Earthjustice, for example, is essentially a mainstream strategy–oriented NGO, whereas the Sierra Club emphasizes participatory strategies, although it applies mainstream approaches too. And the Sierra Club, notably, is the only NGO in this group that engages in 501(c)4 lobbying. Conservation International and World Wildlife Federation represent NGOs that combine mainstream and participatory approaches in roughly equal portions. In terms of the actual number of strategies applied, the large international NGOs generally employ a multitude of strategies. Examples include WWF, CI, and the somewhat smaller entity, Environmental Defense. Others such as TNC and Earthwatch carve out specific niches. TNC emphasizes scientific research and property management whereas Earthwatch emphasizes scientific research and community education. Thus, while all international biodiversity-focused NGOs were not included in this study, the above five organizational characteristics demonstrate that a unique mix of NGOs were examined.

Appendix B
NGO contact information

Conservation International
1919 M Street, NW Suite 600
Washington, DC 20036
Phone: 202-912-1000 or 1-800-406-2306
http://www.conservation.org/

Defenders of Wildlife
1130 17th Street, NW
Washington, DC 20036
Phone: 202-682-9400
http://www.defenders.org

Earthjustice Legal Defense Fund
National Headquarters
426 17th Street, 6th Floor
Oakland, CA 94612-2820
Phone: 510-550-6700
Fax: 510-550-6740
http://www.earthjustice.org/

DC Regional Office
1625 Massachusetts Avenue, NW
Suite 702
Washington, DC 20036
Phone: 202-667-4500

Earthwatch Institute
3 Clock Tower Place
Suite 100
Box 75
Maynard, MA 01754
Phone: 1-800-776-0188 or 978-461-0081
Fax: 978-461-2332
www.earthwatch.org

Environmental Defense
National Headquarters
257 Park Avenue South
New York, NY 10010
Phone: 212-505-2100
Fax: 212-505-2375
    http://www.environmentaldefense.org

Capitol Office
1875 Connecticut Avenue, NW
Washington, DC 20009
Phone: 202-387-3500
Fax: 202-234-6049

The Nature Conservancy
4245 North Fairfax Drive
Suite 100
Arlington, VA 22203-1606
Phone: 703-841-5300
    http://nature.org/

The Ocean Conservancy
1725 DeSales Street, NW
Suite 600
Washington, DC 20036
Phone: 202-429-5609
Fax: 202-872-0619
    http://www.oceanconservancy.org/

The Sierra Club
National Headquarters
85 Second Street, 2nd Floor
San Francisco, CA 94105
Phone: 415-977-5500
Fax: 415-977-5799
    http://www.sierraclub.org/

Legislative Office
408 C Street, NE
Washington, DC 20002
Phone: 202-547-1141
Fax: 202-547-6009

World Resources Institute
10 G Street NE
8th Floor
Washington, DC 20002
Phone: 202-729-7600
Fax: 202-729-7610
http://www.wri.org/

World Wildlife Fund
WWF-US
1250 24th Street, NW
Washington, DC 20037-1193
Phone: 202-293-4800
Fax: 202-293-9211
http://www.worldwildlife.org/

WWF-International
Avenue du Mont-Blanc
1196 Gland
Switzerland
Phone: +41 22 364 91 11
Fax: +41 22 364 53 58
http://www.wwf.org/ or http://www.panda.org/

Notes

INTRODUCTION *(pages 1–10)*

1. Noel L. Grove, "Quietly Conserving Nature," *National Geographic*, 174, no. 6 (December 1988): 824.
2. Thomas Homer-Dixon and Michael Klare are examples of just two scholars who are well known for exploring the ecological link to conflict. For more on this see Thomas Homer-Dixon and Jessica Blitt, eds., *Ecoviolence: Links Among Environment, Population, and Security* (Lanham, Md.: Rowman & Littlefield Publishers, Inc., 1998); and Michael Klare, *Resource Wars: The New Landscape of Conflict* (New York: Metropolitan Books, 2001).
3. Scholars may choose from several terms to discuss interactions across state borders. For years "international" was the term of choice as it literally highlights the interactions "between states." But while the word international will always hold some merit, analyses are increasingly applying the more all-inclusive terms "global" or "transnational," thus paying tribute to the role non-state actors such as NGOs and multinational corporations play on the world stage today.
4. As will be outlined in the NGO subsection of the political setting outlined in chapter 1, the functions of NGOs include scientific or technical expertise, presentation of an alternative power source, and construction as well as maintenance of a global civil society.
5. See Appendix A for a description of case selection here. Also see Appendix B for contact information on each of these environmental NGOs.
6. See chapter 1 for further description of these eleven NGOs. Also, see chapter 4 for a comparison of key organizational characteristics here.
7. Michael P. Cohen, *The History of the Sierra Club: 1892–1970* (San Francisco: Sierra Club Books, 1988); Tom Turner, *Justice on Earth: Earthjustice and the People It Has Served* (White River Junction, Vt.: Chelsea Green Books, 2002); W. William Weeks, *Beyond the Ark: Tools for an Ecosystem Approach to Conservation* (Washington, D.C.: Island Press, 1997); Paul Wapner, *Environmental Activism and World Civic Politics* (Albany: State University of New York Press, 1996).

8. Future research should continue to build in the directions the following chapters outline, targeting other environmental issue areas as well as biodiversity protection. Most notably, there is also a need to focus expressly upon the plethora of Southern Hemisphere NGOs within the developing world, particularly as the Southern Hemisphere is where the vast majority of biodiversity resides. See Figure I.1 for graphic depiction of this fact.

9. Seema Paul, Personal interview with Senior Program Officer, Biodiversity, at United Nations Foundation headquarters, Washington, D.C., October 18, 2002.

10. For a comparison of successful mobilization against the Hidrovia waterway project in the La Plata River basin of Brazil versus the lack of success in Argentina see: Kathryn Hochstetler, "After the Boomerang: Environmental Movements and Politics in the La Plata River Basin," *Global Environmental Politics* 2, no. 4 (November 2002): 35–57.

11. Edward O. Wilson, *The Diversity of Life* (New York: W. W. Norton, 1992). These numbers suggest human activity has increased extinctions by between one and ten thousand times that which would otherwise be expected. Normal "background" extinction as documented in fossil records is approximately one species per one million every year.

12. Joel Tickner, Carolyn Raffensperger, and Nancy Myers, "The Precautionary Principle in Action: A Handbook First Edition," Science and Environmental Health Network, May 31, 2003, <http://www.biotech-info.net/precautionary .html>. The origins of the precautionary principle may be traced to the German principle of *Vorsorge* (foresight), which emphasizes planning in advance to prevent harmful activities on the environment.

13. Joseph Greico, "Anarchy and the Limits of Cooperation: A Realist Critique of the Newest Liberal Institutionalism," *International Organization* 42(August 1988): 485–507.

14. The complete text of the Convention on Biological Diversity is available at several Internet sites. One of these is <http://www.unep.ch/bio/conv-e.html>.

15. Herman E. Daly, *Beyond Growth: The Economics of Sustainable Development* (Boston: Beacon Press, 1996).

16. Herman E. Daly, and John B. Cobb, Jr., *For the Common Good: Redirecting the Economy Toward Community, the Environment, and a Sustainable Future* (Boston: Beacon Press, 1989).

17. While I utilize the research programs of sovereignty and regimes, domestic-international linkage politics, and NGOs, it should also be noted that three grand theories have been applied in interpreting each of these research programs. And each of these theories, namely realism, complex interdependence, and globalization, sheds light on our current predicament. Realism discounts the role of NGOs and emphasizes the primacy of state sovereignty. This theory clearly challenges the significance of NGOs but is also notable for pointing out the obstacles of both power and self-interest that permeate the issue of bio-

diversity protection. Globalization theory takes issue with realism on the significance of NGOs, asserting that states are no longer capable of ensuring stability in the international system and that alternative systems of governance, like the civil society that NGOs support, are needed. Complex interdependence theory lodges itself between these extremes by explaining a world where states still are the primary actors but increasingly turn to non-state entities such as NGOs for assistance in governance.

18. In some respects, the Cold War itself laid the groundwork here as President George Bush's Secretary of State Jim Baker actively sought to foster democracy behind the Iron Curtain by encouraging greater participation of NGOs in central and eastern Europe. NAFTA also created a major schism in the environmental community, as will be discussed later in this book. Some groups such as Greenpeace and the Sierra Club opposed the agreement while WWF, Environmental Defense, NRDC, NWF, and Audubon, to name a few, supported NAFTA.

19. James N. Rosenau, *Linkage Politics: Essays on the Convergence of National and International Systems* (New York: Free Press, 1969).

20. Robert O. Keohane, and Joseph S. Nye, *Power and Interdependence: World Politics in Transition* (Boston: Little, Brown and Company, Inc., 1977).

1. THE TRANSNATIONAL AND INTERDEPENDENT NATURE OF BIODIVERSITY PROTECTION
(pages 11–51)

1. For example, see Richard Falk, *This Endangered Planet: Prospects and Proposals for Human Survival* (New York: Vintage Books, 1971); Al Gore, *Earth in the Balance: Ecology and the Human Spirit* (Boston: Houghton Mifflin Company, 1992); Edward O. Wilson, "The Current State of Biological Diversity," in *Sources: Notable Selections in Environmental Studies,* ed. Theodore D. Goldfarb (Guilford, Conn.: Dushkin Publishing Group, 1997).

2. Edward O. Wilson, *Naturalist* (Washington, D.C.: Island Press, 1994), 359.

3. Yellowstone National Park, "Yellowstone in the Afterglow," June 3, 2003, <http://www.nps.gov/yell/publications/pdfs/firehtmls/chapter1.htm>. The dominant tree in Yellowstone, lodgepole pine also contributes to fires, as its shallow roots are susceptible to strong winds. Over time, as fallen trees and needles gather on the forest floor, they hinder the growth of other leafy plants—until a fire breaks out. The lodgepole pine seedlings benefit from these periodic fires too, as their seedlings cannot survive in the shade of a spruce or fir canopy.

4. Charles Mann, and Mark Plummer, *Noah's Choice: The Future of Endangered Species* (New York: Alfred A. Knopf, 1995).

5. Milton Friedman, *Free to Choose: A Personal Statement* (New York: Harcourt Brace Jovanovich, 1980); Herman Kahn, *The Next 200 Years: A Scenario for America and the World* (New York: Morrow, 1976); Julian Simon, *The Ultimate Resource* (Princeton, N.J.: Princeton University Press, 1981).

6. Francis Bacon, *New Atlantis,* ed. Alfred B. Gough (Oxford: Clarendon Press, 1915); René Descartes, *Discourse on the Method,* ed. David Weissman (New Haven: Yale University Press, 1996).

7. For an in depth examination of this first modern energy crisis, the crisis of deforestation in seventeenth-century Britain, see John U. Nef, "An Early Energy Crisis and Its Consequences," *Scientific American* 237 (November 1977): 140–51.

8. Amory B. Lovins, "Cost-Risk-Benefit Assessments in Energy Policy," *The George Washington Law Review* 45, no. 5 (August 1977): 911–42.

9. Charles J. Glacken, *Traces on the Rhodian Shore: Nature and Culture in Western Thought from Ancient Times to the End of the Eighteenth Century* (Berkeley: University of California Press, 1967).

10. Leo Marx, "Technology: The Emergence of a Hazardous Concept," *Social Research* 64, no. 3 (Fall 1997): 965–88.

11. Henry David Thoreau, *Walden and Other Writings* (New York: Modern Library, 1950).

12. George Perkins Marsh, *Man and Nature,* ed. David Lowenthal (Cambridge: Belknap Press of Harvard University Press, 1965).

13. Wilson, *Naturalist,* 363.

14. In some ways, this is still a chicken-and-egg type of question among biologists, as it is not entirely clear whether speciation precedes or follows the adaptation to a new local environment.

15. Otto T. Solbrig, "The Origin and Function of Biodiversity," *Environment* 33, no. 5 (June 1991): 36.

16. Ibid., 37.

17. Edward O. Wilson, "The Column: Harvard University Press," *New York Times Book Review,* January 14, 1979, 43.

18. World Commission on Environment and Development, *Our Common Future* (Oxford: Oxford University Press, 1987).

19. Herman Daly, and John B. Cobb, Jr., *For the Common Good: Redirecting the Economy Toward Community, the Environment, and a Sustainable Future* (Boston: Beacon Press, 1989).

20. Ernest Yanarella, and Richard S. Levine. "Does Sustainable Development Lead to Sustainability?" *Futures* 18 (October 1992): 759–74.

21. Hazel Henderson, *The Politics of the Solar Age: Alternatives to Economics* (Indianapolis: Knowledge Systems, 1988).

22. James E. Lovelock, *GAIA: A New Look at Life on Earth* (Oxford: Oxford University Press, 1979).

23. Donella Meadows, *The Limits to Growth: A Report for the Club of Rome's Project on the Predicament of Mankind* (New York: Universe Books, 1972).

24. William Ophuls, *Ecology and the Politics of Scarcity* (San Francisco: W.H. Freeman and Company, 1977).

25. Darrell A. Posey, "Protecting Indigenous People's Rights to Biodiversity: People, Property and Bio-prospecting," *Environment* 38, no. 8 (October 1996): 9.

26. Approximately 70 percent of plants identified with cancer-fighting properties live in tropical rainforests.

27. As any student of Irish history knows, the repercussions to relying on one crop can be disastrous. Black rot spread easily through a monoculture during the Irish potato famine of 1846 to 1850 during which over one million people died from hunger and disease. Of course, British political manipulation of food pricing did little to help matters, either.

28. For a comprehensive assessment of this concept see Ken Conca, and Geoffrey D. Dabelko, eds., *Green Planet Blues: Environmental Politics from Stockholm to Kyoto* (Boulder: Westview Press, 1998).

29. Peter H. Raven, "AIBS News: The Politics of Preserving Biodiversity," *BioScience* 40, no. 10 (November 1990): 771.

30. World Resources Institute, "Global Topics: Forests, Grasslands, and Drylands," June 2, 2003, <http://www.wri.org/forests/key_issues.html>.

31. Jeffrey A. McNeely, Kenton R. Miller, Walter V. Reid, Russell A. Mittermeier, and Timothy Werner, "Strategies for Conserving Biodiversity," *Environment* 32, no. 3 (April 1990): 18.

32. Gretchen Daily, ed., *Nature's Services: Societal Dependence on Natural Ecosystems* (Washington, D.C.: Island Press, 1997).

33. Stephen Krasner, "Westphalia and All That," in *Ideas and Foreign Policy: Beliefs, Institutions and Political Change,* ed. Judith Goldstein and Robert Keohane (Ithaca: Cornell University Press, 1993), 235.

34. Andrew Hurrell, "International Society and the Study of Regimes: A Reflective Approach," in *Regime Theory and International Relations,* ed. Volker Rittberger (Oxford: Clarendon Press, 1995), 54.

35. Michael Ross Fowler and Julie Marie Bunck, *Law, Power, and the Sovereign State: The Evolution and Application of the Concept of Sovereignty* (University Park: Pennsylvania State University, 1995), 64.

36. Karen T. Litfin, ed., *The Greening of Sovereignty in World Politics* (Cambridge: The MIT Press, 1998).

37. Krasner, "Westphalia and All That."

38. UN Security Council, "Note by the President of the Security Council," UN Doc. No. S/23500, 1992, p. 3. This raises the question as to what exactly constitutes international security. Security is considered traditionally in military terms—although these conditions often note various environmental and economic factors. Even Hans Morgenthau, the consensus father of modern realism, acknowledges relevant categories such as natural resources, technology, and population in his measures of national power. Still, while these variables were recognized as a factor in international security, Morgenthau did so within the context of how these measures contribute to the military capability of a state. More and more, this peripheral status for environmental issues no longer exists. The United Nations Security Council, for instance, expanded its definition of international security in 1992 to include non-military threats to stabil-

ity. Threats along economic, social, humanitarian, and ecological lines were added.

39. The Treaty of Westphalia ended the Thirty Years War in Europe between shifting coalitions of Protestant states, German principalities, and Catholic France versus the Holy Roman Empire of Habsburg Spain and Austria-Hungary. It is largely regarded as the birth of the modern state system because those who signed the treaty on Prague's famous Charles Bridge represented a degree of legitimacy now found in states instead of the Church.

40. Kenneth Waltz, *Theory of International Politics* (Reading, Mass.: Addison-Wesley, 1979); Robert Gilpin, *War and Change in World Politics* (Cambridge: University Press, 1981).

41. Robert O. Keohane, *Neorealism and Its Critics* (New York: Columbia University Press, 1986); David A. Baldwin, ed., *Neorealism and Neoliberalism: The Contemporary Debate* (New York: Columbia University Press, 1993).

42. Kenichi Ohmae, *The Borderless World: Power and Strategy in the Interlinked Economy* (New York: HarperBusiness, 1990); Marvin S. Soroos, *Beyond Sovereignty: The Challenge of Global Policy* (Columbia: University of South Carolina Press, 1986).

43. Mark W. Zacher, "The Decaying Pillars of the Westphalian Temple: Implications for International Order and Governance," in *Governance Without Government: Order and Change in World Politics,* ed. J. Rosenau and E. O. Czempiel (Cambridge: Cambridge University Press, 1992), 58–101.

44. Joseph Camilleri, "Rethinking Sovereignty in a Shrinking, Fragmented World," in *Contending Sovereignties: Redefining Political Community,* ed. R. B. J. Walker and Saul H. Mendlovitz (Boulder: L. Rienner Publishers, 1990), 38.

45. Oran R. Young, ed., *The Effectiveness of International Environmental Regimes: Causal Connections and Behavioral Mechanisms* (Cambridge: The MIT Press, 1999).

46. Ronnie D. Lipschutz, and Ken Conca, eds., *The State and Social Power in Global Environmental Politics* (New York: Columbia University Press, 1993), xi.

47. Barry B. Hughes, *Continuity and Change in World Politics: The Clash of Perspectives* (Englewood Cliffs, N.J.: Prentice-Hall, 1991), 264.

48. Karen Litfin, "Eco-regimes: Playing Tug of War with the Nation-State," in *The State and Social Power in Global Environmental Politics,* ed. Ronnie D. Lipschutz and Ken Conca (New York: Columbia University Press, 1993), 96.

49. Stevenson Swanson, "Countries Gather at United Nations for Earth Summit," *The Chicago Tribune,* June 26, 1997.

50. Elizabeth Economy, and Miranda A. Schreurs, "Domestic and International Linkages in Environmental Politics," in *The Internationalization of Environmental Protection,* ed. Miranda A. Schreurs and Elizabeth C. Economy (New York: Cambridge University Press, 1997), 15.

51. Robert Putnam posits, for example, a second image explanation in his two-level games, one where domestic variables create international effects. (Robert

Putnam, "Diplomacy and Domestic Politics: The Logic of Two Level Games," *International Organization* 42, no. 3 (Summer 1988): 427–60.) Others such as Peter Gourevitch assert that the second image is actually "reversed," that domestic variation is the result of international causes instead of vice versa. (Peter Gourevitch, "The Second Image Reversed: The International Sources of Domestic Politics," *International Organization* 32, no. 4 (Autumn 1978): 881–911). Christopher G. Thorne, *Border Crossings: Studies in International History* (New York: Basil Blackwell, 1988), 125.

52. Categorization of these four areas of origin draws upon Karen Mingst's "Uncovering the Missing Links: Linkage Actors and their Strategies in Foreign Policy Analysis," in *Foreign Policy Analysis: Continuity and Change in Its Second Generation,* ed. Laura Neack, Jeanne A.K. Hey, and Patrick J. Haney (Englewood Cliffs, N.J.: Prentice Hall, 1995).

53. James N. Rosenau, *Linkage Politics: Essays on the Convergence of National and International Systems* (New York: Free Press, 1969), 10–14.

54. Robert O. Keohane and Joseph S. Nye, *Power and Interdependence: World Politics in Transition* (Boston: Little, Brown and Company, 1977), 24–29.

55. Peter Gourevitch, *Politics in Hard Times: Comparative Responses to International Economic Crises* (Ithaca: Cornell University Press, 1986).

56. Ronald Rogowski, *Commerce and Coalitions: How Trade Affects Domestic Politics Alignments* (Princeton: Princeton University Press, 1989).

57. Marian A. L. Miller, *The Third World in Global Environmental Politics* (Boulder: Lynne Rienner, 1995).

58. Andre Gunder Frank, *Capitalism and Underdevelopment in Latin America: Historical Studies of Chile and Brazil* (New York: Monthly Review Press, 1967).

59. James N. Rosenau, "Governance, Order and Change in World Politics," in *Governance without Government,* ed. Rosenau and Czempiel.

60. Caroline Thomas, ed., *Rio: Unraveling the Consequences* (London: Frank Cass, 1996).

61. Ulrich Beck, "Comments from Environmental Scholars," *Organization & Environment: International Journal for Eco-social Research,* homepage. Available January 11, 2003, at <http://www.coba.usf.edu/jermier/journal.htm#Scholars>.

62. Gareth Porter, and Janet Welsh Brown, *Global Environmental Politics* (Boulder: Westview Press, 1996).

63. Resolution 288 (X) of February 27, 1950, as quoted in Appendix 2, in *Yearbook of International Organizations, 1998/99,* ed. Union of International Associations (Brussels: K.G. Saur, 1999), 1749.

64. For more on the official UN definition, see Jens Martens, "NGOs in the UN System: The Participation of Non-Governmental Organizations in Environment and Development Institutions of the United Nations," *Projectstelle UNCED,* DNR/BUND, Bonn, September 1992, p. 4.

65. Even a cursory review of a number of related directories illustrates this point. For example, see the *World Directory of Environmental Organizations, Inter-*

national Directory of Non-Governmental Organizations, and *Who is Who in Service to the Earth.*

66. Paul Wapner, *Environmental Activism and World Civic Politics* (Albany: State University of New York Press, 1996), 2.

67. Peter Haas, *Saving the Mediterranean: The Politics of International Environmental Cooperation* (New York: Columbia University Press, 1990).

68. Barbara J. Bramble, and Gareth Porter. "Non-Governmental Organizations and the Making of U.S. International Environmental Policy," in *The International Politics of the Environment,* ed. Andrew Hurrell and Benedict Kingsbury (Oxford: Clarendon Press, 1992), 313–53. See especially page 320.

69. Lamont C. Hempel, *Environmental Governance: The Global Challenge* (Washington, D.C.: Island Press, 1996), 144.

70. Alison Jolly, "The Madagascar Challenge: Human Needs and Fragile Ecosystems," in *Environment and the Poor: Development Strategies for a Common Agenda,* ed. H. Jeffrey Leonard (New Brunswick, N.J.: Transaction Books, 1989), 214.

71. Known as the World Wildlife Fund at this time, WWF still holds this name in the United States and Canada. Elsewhere, WWF is known as the World Wide Fund for Nature.

72. Mostafa K. Tolba, with Iwona Rummel-Bulska, *Global Environmental Diplomacy: Negotiating Environmental Agreements for the World, 1973–1992* (Cambridge: The MIT Press, 1998).

73. Ranee K. L. Panjabi, *The Earth Summit at Rio: Politics, Economics, and the Environment* (Boston: Northeastern University Press, 1997), 289.

74. Some now question this neutrality, though, suggesting that NGOs operate like many domestic bureaucratic entities by acquiring a degree of institutional inertia that delegitimizes previous claims toward non-parochial perspectives.

75. Porter and Brown, *Global Environmental Politics,* 54.

76. Porter and Brown cite a March 1990 interview with Barbara Bramble, executive director of the National Wildlife Federation, as their source here.

77. Thomas Princen, Thomas and Mathias Finger, *Environmental NGOs in World Politics: Linking the Local and the Global* (London: Routledge, 1994), 41.

78. Robert Gilpin, "Three Models of the Future," in *Transnational Corporations and World Order,* ed. George Modelski (San Francisco: W.H. Freeman, 1979).

79. Raymond Vernon, *Sovereignty at Bay* (New York: Basic Books, 1971).

80. Jessica T. Mathews, "Power Shift: The Rise of Global Civil Society," *Foreign Affairs* 76 (January/February 1997): 50–66.

81. Lorraine Elliott, *The Global Politics of the Environment* (New York: New York University Press, 1998), 131, 143.

82. Inge Kaul, Isabelle Grunberg, and Marc A. Stern, eds., *Global Public Goods: International Cooperation in the 21st Century* (New York: Oxford University Press, 1999).

83. John McCormick, "The Role of Environmental NGOs," in *The Global Environment: Institutions, Law, and Policy,* ed. Norman J. Vig, and Regina S. Axelrod (Washington, D.C.: Congressional Quarterly Press, 1999), 53.

84. Bruce Rich, *Mortgaging the Earth: The World Bank, Environmental Impoverishment and the Crisis of Development* (Boston: Beacon Press, 1994), 288.

85. Rich, *Mortgaging the Earth,* 289.

86. Rich defines sustainability as "living and working in the present in such a way as to conserve and create as many future options as possible" (ibid., 317).

87. Herman Daly, himself a former World Bank economist, concurs, asserting that growth always reaches a point at which it begins to erode more value than it creates.

88. Wapner, *Environmental Activism.*

89. Sheila Jasanoff, "NGOs and the Environment: From Knowledge to Action," *Third World Quarterly* 18, no. 3 (December 1997): 579–94; and again in conversation at the annual Association for Politics and the Life Sciences convention, Boston, Massachusetts, September 1998.

90. Also of note is the 1989 Basel Convention on the Control of Transboundary Movements of Hazardous Wastes and their Disposal. Like the CBD, this convention targets assistance of developing countries as a primary objective. Unlike the CBD, though, it fails to address broader issues of diversity protection.

91. This is available at <http://www.biodiv.org/convention/partners-websites.asp>.

92. In the United States, under the Endangered Species Act, the U.S. Fish and Wildlife Service (FWS) serves as the primary implementing agency.

93. COP-13 is scheduled for the end of 2004 or in the first half of 2005 in Thailand.

94. Kal Raustiala, "States, NGOs, and International Environmental Institutions," *International Studies Quarterly* 41, no. 4 (December 1997): 719–40.

95. John Lanchbery, "Long-Term Trends in Systems for Implementation Review in International Agreements on Fauna and Flora," in *The Implementation and Effectiveness of International Environmental Commitments: Theory and Practice,* ed. David G. Victor, Kal Raustiala, and Eugene B. Skolnikoff (Cambridge: The MIT Press, 1998), 57–87. The CBD was not included in his analysis because Lanchbery eliminated all agreements reached after 1990 due to "insufficient history of development" of their systems for implementation review.

96. Christine Poupon, "NGOs Aim to Influence UN on Environment and Development," *CERES* 24, no. 2 (March–April 1992): 6.

97. As of June 2003, the last state to enter the treaty as a fully ratified party was Tuvalu on December 20, 2002.

98. Kal Raustiala, and David G. Victor, "Biodiversity Since Rio: The Future of the Convention on Biological Diversity," *Environment* 38, no. 4 (May 1996): 20.

99. The 800-page Agenda 21 document is *non-binding* and considered a blueprint for the Rio principles at large. The Rio Declaration is a *general statement of agreement* that connects environmental and economic concerns. The State-

ment of Forest Principles *recommends* methods for protecting forests from development damage. The Climate Change Treaty calls for *voluntary* reductions of carbon dioxide and other greenhouse gas emissions that are believed to cause global warming.

100. The Conferences of Parties include COP-1 in Nassau, Bahamas (December 1994), COP-2 in Jakarta, Indonesia (November 1995), COP-3 in Buenos Aires, Argentina (November 1996), COP-4 in Bratislava, Slovakia (May 1998), COP-5 in Nairobi, Kenya (May 2000), and COP-6 in The Hague, Netherlands (April 2002). COP-7 was held in Kuala Lumpur, Malaysia, February 9–20, 2004.

101. George W. Bush, "Text of a Letter from the President to Senators Hagel, Helms, Craig, and Roberts," The White House, March 13, 2001, <http://www.whitehouse.gov/news/releases/2001/03/20010314.html>.

102. "Convention on Biological Diversity," June 8, 2000, <http://www.unep.ch/bio/conv-e.html>, p. 11.

103. Ibid., pp. 11–13.

104. As Robert Paarlberg states more generally, "[U.S.] congressional foot dragging poses a greater challenge than differences among nations to U.S. environmental policy leadership." Robert L. Paarlberg, "A Domestic Dispute: Clinton, Congress and International Environmental Policy," *Environment* 38, no. 8 (October 1996): 18.

105. As noted above, partisan politics is a large part of this story. One could make a case for this in the form of several wise-use groups such as the National Federal Lands Conference, Alliance for America, and American Farm Bureau. Perhaps the most intriguing opposition to the CBD of all, though, is that led by convicted felon Lyndon LaRouche. LaRouche claimed the Convention was a new religion and a threat to United States sovereignty as well as basic private property rights. Zionism, the British royal family, drug trade, and UN imperialism all wrapped together to form LaRouche's conspiracy theory. For more discussion on this see Michel W. Robbins, "Biodiversity and Strange Bedfellows" (editorial), *Audubon* 97, no. 1 (January–February 1995): 4.

106. Oran R. Young, and Marc A. Levy. "The Effectiveness of International Environmental Regimes," in *The Effectiveness of International Environmental Regimes: Causal Connections and Behavioral Mechanisms,* ed. Oran Young (Cambridge: The MIT Press, 1999), 3.

107. Oran R. Young, "Regime Effectiveness: Taking Stock," in *The Effectiveness of International Environmental Regimes,* 277.

108. Carroll Muffett, Personal interview with International Programs director, Defenders of Wildlife headquarters, Washington, D.C., October 17, 2002.

109. Young and Levy, "The Effectiveness of International Environmental Regimes," 4.

110. Ibid., 6.

111. David G. Victor, Kal Raustiala, and Eugene B. Skoknikoff, *The Implementation and Effectiveness of International Environmental Commitments: Theory and Practice* (Cambridge: The MIT Press, 1998), 6–7.

112. Peter M. Haas, Robert O. Keohane, and Marc A. Levy, *Institutions for the Earth: Sources of Effective International Environmental Protection* (Cambridge: The MIT Press, 1994), 10–24.

113. For a discussion of how regimes allocate various roles to participants see Alexander Wendt, "Anarchy is What States Make of It: The Social Construction of Power Politics," *International Organization* 46 (1992): 391–425.

114. Donald T. Wells, *Environmental Policy: A Global Perspective for the Twenty-First Century* (Upper Saddle River, N.J.: Prentice-Hall, Inc., 1996).

115. Scholars such as David Victor, Kal Raustiala, and Eugene Skoknikoff emphasize implementation throughout their edited volume of fourteen case studies covering eight major areas of international environmental regulation.

116. This research examines the ability of environmental NGOs to make three fundamental linkages by changing behavior. Although elements of what constitutes the necessary conditions for success are implicit in this discussion, I do not make an outright case for the sufficient conditions to achieve policy success, as NGOs can be effective without being, at least immediately, successful. See Young and Levy, "The Effectiveness of International Environmental Regimes," 19–28.

117. James Coleman, *Foundations of Social Theory* (Cambridge: Harvard University, 1990). See also Robert Putnam's oft-cited *Making Democracy Work: Civic Traditions in Modern Italy* (Princeton, N.J.: Princeton University Press, 1993).

118. For further elaboration see Dave Foreman, *Confessions of an Eco-Warrior* (New York: Harmony Books, 1991).

119. An excellent summary of the three types of funding (non-tax-deductible, tax-deductible, and electioneering), as well as the critical distinctions between 501(c)3 and 501(c)4 money is found in the Sierra Club's chapter on "Managing Your Money" in their *Grassroots Organizing Training Manual* (San Francisco: Sierra Club, 1999).

120. Edward Abbey, *The Monkey Wrench Gang* (New York: Avon Books, 1975).

121. From Appendix 2, "Types of Organizations," Union of International Associations, ed., *Yearbook of International Organizations, 1998/99,* vol. 3 (Brussels: KG Saur, 1999), 1749–50.

122. Gary C. Bryner, *From Promises to Performance: Achieving Global Environmental Goals* (New York: W. W. Norton & Company, 1997), 209.

123. Statistics reported by the American Institute of Philanthropy, a nonprofit charity watchdog that publishes a tri-annual *Charity Watchdog Report,* and GuideStar, a national database of more than 850,000 IRS-recognized nonprofit organizations, allow measurement here.

124. Karen Mingst, "Implementing International Environmental Treaties: The

Role of NGOs," paper presented at the annual meeting of the International Studies Association, Mexico, 1993.

125. Bramble and Porter, "Non-Governmental Organizations," 348.

2. WORKING WITHIN THE SYSTEM *(pages 52–94)*

1. Mindy Kay Bricker, "*Time* [European edition] Names Local Scientist a 'Hero': Hydrologist Josef Krecek Uncovered Environmental Damage in North Bohemia," *The Prague Post,* May 15, 2003.
2. Time Europe Magazine, "Heroes 2003," *Time Europe,* April 28, 2003, <http://www.time.com/time/europe/hero/index.html>. Thirty-six Europeans were honored in categories ranging from two others in Krecek's "green team" division to activists, hate busters, and innovators.
3. Maryann Bird, "A Light in the Black Triangle," *Time Europe* 161, no. 17 (April 28, 2003).
4. Earthwatch Institute, *2002–2003 Research and Exploration* 21, no. 3: 54. <http://www.time.com/time/europe/magazine/printout/0,13155,901030428-442027,00.html>.
5. Joseph Krecek, Associate Professor Lecturer, Institute of Natural Resources, European University Department of Hydrobiology, Letter to Mountain Waters of Bohemia prospective participants, Prague, Czech Republic, November 23, 2001, reproduced in Earthwatch Institute's "Meet the Scientists," <http://www.earthwatch.org/expeditions/krecek/meetthescientists.html#staff>.
6. For a discussion of the enhanced professionalism of lobbyists in general see David Segal, "A Nation of Lobbyists," *The Washington Post National Weekly Edition* 12, no. 37 (1995): 11.
7. The word "the" in The Nature Conservancy is capitalized in this book in deference to that NGO's practice of capitalizing it in its literature and in the inclusion of this word in its popularly referred to acronym of TNC. This stands in contrast to NGOs such as the Sierra Club and the Ocean Conservancy, neither of which use this article in their acronym and rarely capitalize the article "the" in referring to themselves in their literature.
8. NIMBY stands for "not in my backyard."
9. Ian A. Bowles and Cyril F. Kormos, "Environmental Reform at the World Bank: The Role of the U.S. Congress," *Virginia Journal of International Law* 35, no. 4 (Summer 1995): 777–839.
10. Cyril Kormos, personal interview at Conservation International headquarters, Washington, D.C., October 20, 1999.
11. Bruce Rich (personal interview, October 22, 1999), for one, believes that the World Bank still requires significant revamping and even questions whether energy invested is worthwhile, suggesting such efforts might be better spent elsewhere.
12. Conservation International, "Creating Solutions for the 21st Century: 1998 Annual Report."

13. James Sheehan and Paul Georgia, "Feeding the Green Money Tree," *The Washington Times,* July 29, 1999, A-20.
14. *Scenic Hudson Preservation Conference v. Federal Power Commission,* 453 F2nd 463.
15. For an in-depth description of the dolphin-tuna story, see Nina M. Young, Robert W. Irvin, and Meredith L. McLean, "The Flipper Phenomenon: Perspectives on the Panama Declaration and the 'Dolphin Safe' Label," *Ocean and Coastal Law Journal* 3 (1997): 57–115.
16. Bob Irvin, personal interview with Vice President of Marine Wildlife Conservation and General Counsel at the Ocean Conservancy's headquarters in Washington, D.C., October 19, 1999.
17. Earthjustice Legal Defense Fund, May 30, 1999, <http://www.earthjustice .org/about/index.html>.
18. Earthjustice Legal Defense Fund, *Annual Report 1998,* 12–13.
19. South and Meso American Indian Rights Center (SAIIC), "U'wa of Colombia Reject All New Oil Exploration: New Report Details Occidental Petroleum's Role in Ongoing Crisis," press release, August 10, 1998, <http://www.hartford-hwp.com/archives/41/171.html>.
20. Amazon Watch, Project Underground, and Rainforest Action Network, "Colombia's U'wa Tribe and Supporters Celebrate Oxy's Failure to Find Oil," press release, July 31, 2001, <http://www.amazonwatch.org/newsroom/ newsreleases01/jul31_uwa.html>. A number of social issues beyond ecological devastation were at stake in this case as well. Like many other areas around the world where oil exploration takes place, the added pressures on water supplies and food resources are borne most directly by the local populations. And, time and again, the added infrastructure to support drilling for oil also spurs mass movements of people into what were previously relatively isolated territories. Innumerable undesirable side effects develop. Crime and prostitution invariably increase, for example.
21. Rainforest Action Network, Network Native Forest Network, and ACERCA (Action for Community and Ecology in the Rainforests of Central America) took issue with the $500,000 in Occidental stock that Gore had inherited from his father, who was on their board of directors, and the fact that Occidental contributed to the Clinton and Gore political campaigns. Five protestors were arrested for criminally trespassing and resisting arrest.
22. OXY received drilling rights to the Siriri block, formerly known as Samoré, in 1992.
23. Sierra Legal Defence Fund, "About Us." June 3, 2003, <http://www.sierralegal .org/aboutsierralegal.html>.
24. The English language Web site is available at <http://www.aida2.org/english/ index.php> . A Spanish Web site is also available.
25. Earthjustice Legal Defense Fund, "Docket 1999."
26. Not to be confused with Santa Barbara–based Environmental Defense Center,

which is also a nonprofit law firm, but one that serves primarily California's
south-central coast since its founding in 1977.

27. For the next five years, on a state-by-state basis within the United States, Environmental Defense continued the fight until winning a ban on DDT consumption domestically in 1972. DDT remains in use internationally, most often in efforts to control malaria.

28. Environmental Defense Fund, "EDF's International Program at Work" (undated pamphlet).

29. Anne O'Brien, "Purchasing Power: Why We *Still* Buy Land," *Nature Conservancy* (November/December 1999), 12–17.

30. As will be discussed in greater detail in chapter 4, respect does not necessarily translate into admiration. For example, although they do work on many issues together, there remains some residual bitterness between The Nature Conservancy and Conservation International after former international program members in TNC left that organization and founded CI in January 1987. Interestingly, the disagreement largely was over emphasis, or more accurately lack thereof, upon TNC's international agenda.

31. Thomas Lovejoy, "Aid Debtor Nation's Ecology," *New York Times,* October 4, 1984.

32. At the first Intersessional meeting on Operations of the Convention on Biological Diversity (ISOC-1), for example, this author was classified as an NGO from the University of Kentucky, given the respective pink color-coded badge, and seated with my "fellow" NGOs at the back of the assembly hall for the Plenary Session each day. Held June 28–30, 1999, at the International Civil Aviation Organization in Montreal, Canada, daily events also included both morning and afternoon strategy sessions of those NGOs present.

33. In 1998, 20 percent of TNC's budget was allocated to its international program.

34. Officials at TNC recognize the difficulty inherent in selecting an area big enough to have an impact on the size of a problem. They fear that many biodiversity protection initiatives are unfortunately disproportionately small to the size of the issue.

35. USAID spent what might first appear to be a whopping $68 million on global biodiversity conservation in fiscal year 1999. But this must be framed in the context of a U.S. foreign operations budget (of which USAID is a part) totaling over $12 billion.

36. Former TNC President and CEO John Sawhill outlines this philosophy in John C. Sawhill, "The Nature Conservancy," *Environment* 38, no. 5 (June 1996): 43–44.

37. Reshma Prakash, personal interview at *The Earth Times* office in New York City, August 3, 1999. Founded in 1991, *The Earth Times* prides itself as the newspaper of record for the environmental community since Rio, where it circulated some 35,000 print copies daily. The paper pays particular attention to the United Nations system as well as to bilateral and multilateral lending institutions. It is published daily on the Internet, and two times a month (1st and

15th) in print by the not-for-profit International Media Earth Times Foundation. According to its mission statement, the target audience is policymakers, community and business leaders, NGOs in both developed and developing states, and students and teachers worldwide.

38. Bob Irvin, personal interview at the Ocean Conservancy headquarters, Washington, D.C., October 19, 1999.

39. The Ocean Conservancy, though, does not disclaim conscientious protest positions from time to time. Their participation in the International Whaling Commission is a case in point. From their ethical stance grounded in emotive values, not science, the Ocean Conservancy actively opposes any and all whaling. This is actually a bit ironic considering their stance on the dolphin-tuna debate.

40. David Guggenheim, personal interview with Vice President for Conservation Policy at the Ocean Conservancy headquarters, Washington, D.C., October 17, 2002.

41. Alan Patterson, "Debt for Nature Swaps: And the Need for Alternatives," *Environment* 32, no. 10 (December 1990): 5–13, 31–32.

42. As much as $75 million in matching funds is included in the grant, going toward CI's Global Conservation Fund.

43. Keith Alger, personal interview with Vice President, Conservation Strategy Department, Conservation International headquarters, Washington, D.C., October 17, 2002.

44. Ten such RAP working papers are available through the Conservation International Web site at <www.conservation.org/WEBfieldact/c-c_prog/science/history.htm>.

45. Cyril Kormos, personal interview with Research Associate at Conservation International headquarters, Washington, D.C., October 20, 1999.

46. See chapter 4 for further discussion of this partnership phenomenon.

47. Earthjustice Legal Defense Fund, "Docket 2000."

48. Kathryn S. Fuller, "President's Message: Thinking and Acting Globally and Locally," *Focus* 22, no. 1 (January/February 2000), 2.

49. Bruce Rich, personal interview at Environmental Defense Fund's Capitol Hill office, Washington, D.C., October 22, 1999.

50. Nicholas Lapham, personal interview with Program Officer for Environment, United Nations Foundation, Washington, D.C., July 30, 1999.

51. Franklin Moore, personal interview at the United States Agency for International Development, Washington, D.C., October 21, 1999.

52. Sharon Beder, *Global Spin: The Corporate Assault on Environmentalism* (White River Junction, Vt.: Chelsea Green Publishing Company, 2002), 130–31.

53. Ibid., 273.

54. David B. Ottaway and Joe Stephens, "Nonprofit Land Bank Amasses Billions: Charity Builds Assets on Corporate Partnerships," *The Washington Post,* May 4, 2003, A-1.

55. Jim Hightower, "Get the Hogs Out of the Creek!" *Earth Island Journal* 11, no. 1 (1995): 32.

56. Defenders offers three blends through its roaster, Thanksgiving Coffee Company: light, dark, and decaf.
57. Martin Wolk, "Starbucks Hops on Bandwagon with Eco-Friendly Coffee," *Planet Ark,* August 4, 1999, <http://www.planetark.org/dailynewsstory.cfm?newsid=2784&newsdate=04-Ag-1999.htm>.
58. This is also notable to the extent that these small plots act as buffer zones in taking pressure off protected areas.
59. Environmental Defense Fund, "Who We Are, What We Do, How You Can Help: A Guide for Members and Friends," back cover of pamphlet.
60. Michael Bean, personal interview with Director of Endangered Species and Wildlife, Environmental Defense, Washington, D.C. Congressional office, July 26, 1999.
61. Tom Turner, telephone interview with Senior Editor, Earthjustice, Oakland, California, June 2, 2003.
62. Conservation International, "Promoting Conservation Through International Assistance," June 25, 2000, <http://www.conservation.org/WEB/FIELDACT/C-C_PROG.policy.intlasst.htm>.
63. WWF Global Network, "Dams Are Direct Cause of Species Decline, says WWF," April 5, 2000, <http://www.panda.org/news/press/news.cfm?id=1910>.
64. As we shall see in chapter 3, this strategy is particularly effective when it accompanies the participatory strategy of educational initiatives.
65. Green Earth Organization et al., "Statement of the Fourteenth Global Biodiversity Forum," June 18–20, 1999, Montreal, Quebec, Canada.
66. The steering committee consisted of representatives from member organizations and a smaller executive committee from this group was responsible for day-to-day decisions, including financial and administrative matters. Worldwatch Institute, The Nature Conservancy, the Center for International Environmental Law, Environmental Defense, World Resources Institute, Healing Forest Conservancy, Conservation International, Defenders of Wildlife, and the World Conservation Union were all members of the steering committee.
67. Hans Verolme, personal interview with Coordinator of Biodiversity Action Network (BIONET), Washington, D.C., headquarters, July 28, 1999.
68. For example, the Jakarta Mandate on Marine and Coastal Biodiversity was highly confusing, so BIONET created a user-friendly guide for states.
69. "The GEF in the 21st Century: A Vision for Strengthening the Global Environment Facility," December 20, 1999, <htttp://www.igc.org/bionet/gef21.html> [URL no longer available].
70. The Ocean Conservancy, "National Marine Debris Monitoring Program Expands to the West Coast," *1998 Annual Report,* 5.
71. Conservation International, "Monitoring and Evaluation Program: Performance and Management," June 25, 2000, <http://www.conservation.org/WEB/ABOUTCI/Monitor.htm>.
72. Mark W. Zacher, "The Decaying Pillars of the Westphalian Temple: Implica-

tions for International Order and Governance," *Governance Without Government: Order and Change in World Politics,* ed. J. Rosenau and E. O. Czempiel, (Cambridge: Cambridge University Press, 1992), 58–101.

73. Estraleta Fitzhugh, personal interview with Department of Government Relations Liaison, World Wildlife Fund, Washington, D.C., July 30, 1999.
74. Capacity building is the new catchphrase in both environmental and developmental circles, but many increasingly fear that this term is too bureaucratic. Democracy governance has emerged as a viable alternative.
75. Lisa Speckhardt, "Litigation—An Essential Tool for Environmental Protection," *EarthFocus: Friends of the Earth's News Magazine* 30, no. 1 (Spring 2000): 4–5.
76. Also note here the symbiosis between mainstream and participatory strategies.
77. O'Brien, "Purchasing Power," 17.

3. WORKING WITH PEOPLE: PARTICIPATORY STRATEGIES *(pages 95–138)*

1. Michael P. Cohen, *The History of the Sierra Club: 1892–1970* (San Francisco: Sierra Club Books, 1988), 163.
2. According to Brower in an interview with then–Sierra Club president Adam Werbach, the "druid" label originated with a developer who wanted to develop Cumberland Island, while biographer John McPhee added the "arch." Adam Werbach, *Act Now, Apologize Later* (New York: HarperCollins, 1997), 287.
3. Sections of each advertisement were set off as coupons that the reader could cut out and mail to the Sierra Club requesting membership as well as memoranda to send their representatives, senators, Secretary of Interior Udall, and even President Lyndon Johnson to protest destruction of the Grand Canyon.
4. Cohen, *The History of the Sierra Club,* 362.
5. The mainstream strategy of scientific research is critical here in terms of gathering data, of course, but to reach the general public the data must pass through participatory strategies.
6. This seemingly anthropocentric perspective carries greater sway with the general public than it would with specialists in ecology, and it does not necessarily betray ecological thought if one considers the degree to which interdependence makes the human-environmental boundary a permeable one.
7. The potential of combining strategies also underscores the rationale behind my analyzing effectiveness toward meeting the three fundamental linkages rather than specific strategies.
8. For more on Sierra Club's recreational programs as well as a growing list of international, tourist-type events, see the Sierra Club Web pages themselves at <http://www.sierraclub.org/outings/>.
9. This trip ranges from the southern tip of the Cape of Good Hope near Cape Town to the northern edge of South Africa with a safari in Kruger National Park. An in-depth description is found on the Web at <http://www.sierraclub.org/outings/national/copy/00640b.asp>.

10. An outline of each of Sierra Club's initiatives can be found at <http://www
.sierraclub.org/environment/globalissues.asp>.

11. Global warming received special program designation in the year 2000, and
remains one of the highlighted issues that members may access from the Sierra
Club home page. From here, a series of pages on the Web site explore the various relationships within this complex phenomenon. For more information see
<http://www.sierraclub.org/globalwarming/café/factsheet/wildlife.asp>.

12. Note how the focus of the photograph is the eyes of a child, thus depicting future needs and touching upon a common sense of responsibility.

13. The Ocean Conservancy's International Coastal Cleanup serves as a good model
here. This is particularly true as it fosters future contacts among its volunteers,
contacts that cross not only local but also national political boundaries.

14. Rodger Schlickeisen, "The Internet Revolution," *Defenders,* Spring 2000,
<http://www.defenders.org/magazinenew/rs/rssp00.html>.

15. The Sierra Club also offers a series of e-mail lists for members to keep in contact with one another. These are used primarily for local, U.S. state, or national
issues in the United States.

16. Archives of EDF dispatch are available at <http://plaza.edf.org/dispatches.nsf>.

17. Environmental Defense, "Environmental Defense Action Network 1999 Wrap
Up," <EDF@actionnetwork.org> listerv, February 4, 2000.

18. The Web address is <http://www.actionnetwork.org/>.

19. From 1995 to 2000, Defenders also sponsored *GREENlines,* a free weekday
e-mail newsletter of GREEN, Defenders' Grassroots Environmental Effectiveness Network. This is not to be confused with *Greenwire,* an independently distributed, electronic, and daily briefing on environmental news and events that
is still in operation. Covering news at the state, national, and international
level, Greenwire synthesizes the coverage of more than two hundred newspapers and broadcast networks.

20. Defenders of Wildlife, "DENlines Issue #13 (Extended Earth Day Issue)," on-
line posting, discussion listserv, available at <http://www.denaction.org/>,
April 21, 2000.

21. Defenders includes a list of names and addresses of major U.S. grocery store
chains that have committed to dolphin-safe tuna as well as those that have not.
These two lists, a copy of the form letter, and several fact sheets on the dolphin-
safe tuna issue are posted on the Defenders Web site at <www.defenders.org/
wildlife/dolphin/>.

22. Defenders Electronic Network (DEN) is available at <http://www.denaction
.org/>.

23. Founded in 1997, the Web Marketing Association (WMA) rates marketing and
corporate Web development on the World Wide Web in seventy-eight industry categories. It is staffed by volunteers from the Internet world of marketing,
advertising, public relations, and design.

24. World Wildlife Fund, "World Wildlife Fund at Work," June 1999, 12. Rhi-

noceros horns are used in traditional Asian medicines and as dagger handles in the Middle East. Tiger parts are also used for various medicinal purposes as well as in rugs, trophies, and good luck charms. Elephants are used for their ivory and meat—and also killed simply because they are a competitor for land with burgeoning human populations in Africa and Southeast Asia.

25. Nearly three thousand messages to leaders around the world urged more effective conservation and management of sharks, which are killed for their fins and other parts.

26. WWF Conservation Action Network, "Mexico Cancels Salt Works Planned for Pristine Ecoregion of Importance to Gray Whales," March 15, 2000, results archive, <http://takeaction.worldwildlife.org/results/thanks.asp>.

27. Available at <http://takeaction.worldwildlife.org/>.

28. Dennis Cosgrove, "Contested Global Visions: One-World, Whole Earth, and the Apollo Space Program," *Annals of the Association of American Geographers* 84, no. 2: 270–94.

29. Not all NGOs have a membership, of course. This refers only to those NGOs that do have members.

30. Ernest J. Yanarella, "Environmental vs. Ecological Perspectives on Acid Rain: The American Environmental Movement and the West German Green Party," in *The Acid Rain Debate: Scientific, Economic, and Political Dimensions,* ed. Ernest J. Yanarella and Randall H. Ihara (Boulder: Westview Press, 1985), 245–46; Sheldon Wolin, "Editorial," *Democracy* 2 (July 1982): 2–4.

31. These issue advertisements comply with federal election regulations in that they do not expressly endorse candidates. Richard L. Berke, "Sierra Club Ads in Political Races Offer a Case of 'Issue Advocacy,'" *The New York Times,* October 24, 1999, A-12.

32. Elizabeth Hagan, "Three Babes and a Bus Named 'Bella:' The Sierra Student Coalition Rainforest Education Bus Mania Swept the East Coast this May!" *Generation E: The Voice of the Sierra Student Coalition,* Summer 1999, 4.

33. Earthwatch Institute, *Earthwatch Institute 1999 Research and Exploration.* Watertown, Mass.: Earthwatch Expedition, Inc. 1999, inside cover page.

34. Blue Magruder, personal interview with the Public Relations Director at Earthwatch Institute headquarters, Watertown, Massachusetts, July 8, 1999.

35. Earthwatch Institute, "The Center for Field Research at Earthwatch Institute Applications Guidelines Summary."

36. This funding tends to be in smaller increments, though, so Earthwatch can continue to afford to be daring.

37. Julianne Basinger, "To Scientists who use Paying Volunteers in Fieldwork, the Benefits Outweigh the Bother," *The Chronicle of Higher Education,* June 19, 1998.

38. On a related note, researchers also comment upon the considerable time and energy expended in constantly having to train people in data-gathering techniques.

39. Two addresses of interest here are Earthwatch's global classroom at <http://www.earthwatch.org/ed/home.html> and its recommended site for resources at <http://www.earthwatch.org/ed/olr/resources.html>.

40. Earthwatch Institute, "How Earthwatch Works," in *Earthwatch Institute 1999 Research and Exploration* (Watertown, Mass.: Earthwatch Expedition, Inc. 1999).

41. Stanley W. Burgiel and Sheldon Cohen, *The Global Environment Facility From Rio to New Delhi: A Guide for NGOs* (Gland, Switzerland: IUCN, 1997).

42. Intersessional Meetings on the Operations of the Convention are set for every two years, falling between Conference of Parties (COPs) to the Convention on Biological Diversity (CBD). They are held in Montreal, the seat of the Secretariat to the CBD.

43. Personal observations at the inaugural Intersessional Meeting on the Operations (ISOC-1) of the Convention on Biological Diversity, at the International Civil Aviation Organization, Montreal, Canada, June 28–30, 1999.

44. Hazel Henderson, *The Politics of the Solar Age: Alternatives to Economics* (Indianapolis: Knowledge Systems, 1988).

45. Dan Seligman, personal interview at the Sierra Club legislative headquarters, Washington, D.C., October 19, 1999.

46. Dan Seligman provided this critique in his explanation of the lack of a viable domestic constituency for transnational biodiversity protection. I believe it is an accurate one as evidenced by the paucity of international programs within the major U.S. environmental NGOs, at least until the 1990s. Environmental NGOs, throughout much of the 1980s and into the 1990s, misjudged the scale and scope of their agenda and were forced into reactionary mode.

47. Tom Turner, telephone interview with Senior Editor, Earthjustice, Oakland, California, June 2, 2003.

48. Of course, the short-term lesson of the second Gulf War is that adamant opposition to the United States plan for Iraq failed to prevent the U.S.-British operation, but perhaps the long-term lessons will be decidedly more multilateral in nature.

49. Stephen Mills, personal interview with Director, International Program, Sierra Club, Washington, D.C., October 18, 2002.

50. These tips are available at <http://www.sierraclub.org/takeaction/toolkit/>.

51. Dan Seligman, personal interview at the Sierra Club legislative headquarters, Washington, D.C., October 19, 1999.

52. The Nature Conservancy, "International Trips 1999."

53. The cost of these trips begins at approximately $2,000 per person. The fee includes accommodations, meals, fees for guides, gratuities, and often airport departure taxes. Some funds are also generally left over for conservation programs, so this tactic qualifies as a fundraising initiative as well as linkage activity. World Wildlife Fund, "Member Tours at World Wildlife Fund—US." E-mail correspondence with author, July 13, 2000.

54. World Wildlife Fund, "World Wildlife Fund Travel," *Focus* 22, no. 2 (March/April 2000), 7.
55. Ibid., 3.
56. Earthwatch Institute, *2002–2003 Research and Exploration* 21, no. 3: 21.
57. Blue Magruder, personal interview at Earthwatch Institute headquarters, Watertown, Massachusetts,, July 8, 1999.
58. Conservation International Annual Report 1998, 26–27.
59. Martha S. Honey, "Treading Lightly? Ecotourism's Impact on the Environment," *Environment* 41, no. 5 (June 1999): 4–9, 28–33.
60. Conservation International, "Film Capturing Guyana's Natural Heritage a Finalist at Jackson Hole," Press Release, September 23, 1999. The Jackson Hole Wildlife Film Festival features environmental pieces biannually in September in Grand Teton National Park outside of Jackson, Wyoming. The festival regularly includes producers and directors from the BBC, National Geographic, and Discovery Channel.
61. Haroldo Castro, personal interview with the vice president of international communications, Conservation International headquarters, Washington, D.C., October 16, 2002.
62. Environmental Defense, "Action Alert," online posting, discussion listserv, June 1, 2000.
63. Korinna Horta, "Rhetoric and Reality: Human Rights and the World Bank," *Harvard Human Rights Journal* 15 (Spring 2002): 227–43.
64. United Nations (Press Release HE/916), "UNEP Releases First Global Biodiversity Assessment Report," November 14, 1995, available on WRI's Web site February 22, 2000, at <www.wri.org/biodiv/gba-unpr.html>.
65. Martha C. Monroe, ed., *What Works: A Guide to Environmental Education and Communication Projects for Practitioners and Donors* (Washington, D.C.: Academy for Educational Development, 1999); Pat Foster-Turley, *Making Biodiversity Happen: The Role of Environmental Education and Communication* (Washington, D.C.: GreenCOM, 1996).
66. As a division in the Academy for Educational Development, GreenCOM serves as a resource for environmental NGOs in the United States and those abroad.
67. Peter Spain, personal interview with the administrative director of the Academy for Educational Development at GreenCOM headquarters, Washington, D.C., October 20, 1999.
68. For more information on each of these projects and a comprehensive list of activities see the Earthwatch Global Classroom pages at <http://www.earthwatch.org/ed/olr/biodiversity.html>.
69. At one point, Environmental Defense even maintained a series of pages devoted to children themselves. Their "Earth to Kids" pages offered poetry contests, games, wildlife art, and a list of children's books as well as a teacher discussion forum.

70. Basinger, "To Scientists Who Use Paying Volunteers in Fieldwork."
71. Earthwatch Institute, "Education Programs at Earthwatch," June 25, 2000, <http://www.earthwatch.org/aboutew/education.html>.
72. World Resources Institute was one of five nominees in the activism category for 2003. ActForChange won the award, with Greenpeace receiving the People's Choice award.
73. World Resources Institute, "WRI Article: Building Biodiversity Awareness in Schools," June 25, 2000, <http://www.wri.org/wri/biodiv/b33-gbs.html>.
74. World Wildlife Fund, "World Wildlife Fund at Work," June 1999, 13.
75. The Center for Marine Conservation, "Center for Marine Conservation—International Coastal Cleanup," June 25, 2000, <http://www.cmc-ocean.org/cleanupbro/about.php3>.
76. Everette E. Dennis, "In Context: Environmentalism in the System of News," in *Media and the Environment,* ed. Craig L. LaMay and Everette E. Dennis (Washington, D.C.: Island Press, 1991), 57.
77. While voiced in one form or another by numerous NGO representatives, this recommendation was perhaps most eloquently stated by Nicholas Lapham, Program Officer for Biodiversity at the United Nations Foundation.
78. This discussion of irreversibility deserves a qualification of sorts. Technically, as world-renowned paleontologist and evolutionary theorist Stephen Jay Gould states, corrections are possible for the Earth itself. These corrections, however, develop over millions of years—not soon enough to make a difference for humans. Gould, Stephen Jay, *Eight Little Piggies: Reflections in Natural History* (New York: W.W. Norton & Company, 1993).
79. Don Lind, "The Earth Home We See from Space," *National Forum* 75, no. 1 (Winter 1995): 15.
80. For more on the psychological effects of astronauts seeing the Earth from space see Frank White, *The Overview Effect* (Boston: Houghton Mifflin, 1987).
81. David J. Elkins and Richard E. B. Simeon, "A Cause in Search of its Effect, or What Does Political Culture Explain," *Comparative Politics* 11 (January 1979): 127–46.

4. WORKING ON THEMSELVES: ORGANIZATIONAL STRUCTURE *(pages 139–77)*

1. Noel Grove, "Quietly Conserving Nature," *National Geographic* 174, no. 6 (December 1988), 844.
2. A number of sources provided this story, several of which must remain anonymous. Perhaps the most unbiased perspective is that relayed by Rod Mast, vice president of CI-Soujourns at Conservation International. At the time of the "palace revolt," Mast was working at WWF with renowned ethno-botanist Mark Plotkin, whose wife, Liliana Madrigal was head of the Costa Rica division at TNC. Madgrigal and Plotkin are an intriguing couple in their own right, going on to found the group ACT (Amazon Conservation Team) in 1995 as

chief operating officer and president, respectively. ACT has certainly carved out its own niche, working with indigenous peoples to preserve their culture as well as their wisdom about the medicinal properties of their surroundings. In this manner ACT marries traditional tribal knowledge with Western science.

3. David Truman, *The Governmental Process* (New York: Alfred A. Knopf, 1951).
4. Seema Paul, personal interview with Senior Program Officer, Biodiversity, at United Nations Foundation headquarters, Washington, D.C., October 18, 2002.
5. John M. (Mick) Seidl, telephone interview with Chief Program Officer, Environment, Gordon and Betty Moore Foundation, San Francisco, California, June 4, 2003.
6. William P. Browne, *Groups, Interests, and U.S. Public Policy* (Washington, D.C.: Georgetown University Press, 1998), 1–8. Browne asserts that three principles guide how policymaking takes place: that organized interests matter, that people within those organized interests matter, and that the issue must be understood within the larger context of policymaking itself.
7. Ronald G. Shaiko, *Voices and Echoes for the Environment: Public Interest Representation in the 1990s and Beyond* (New York: Columbia University Press, 1999), p. xiv.
8. Karen Mingst, "Implementing International Environmental Treaties: The Role of NGOs," paper presented at the annual meeting of the International Studies Association, Mexico, 1993.
9. Leon Gordenker and Thomas G. Weiss, "Pluralizing Global Governance: Analytical Approaches and Dimensions," in *NGOs, the UN, and Global Governance,* ed. Thomas G. Weiss and Leon Gordenker (Boulder: Lynne Rienner Publishers, 1996), 25–30.
10. Ibid., 35.
11. Seema Paul, personal interview with Senior Program Officer, Biodiversity, at United Nations Foundation headquarters, Washington, D.C., October 18, 2002.
12. Russell J. Dalton, *Citizen Politics in Western Democracies: Public Opinion and Political Parties in the United States, Great Britain, West Germany, and France* (Chatham, N.J.: Chatham House Publishers, Inc., 1988), 110.
13. For more on amphibian die-offs, see Kathryn Phillips, *Tracking the Vanishing Frogs: An Ecological Mystery* (New York: St. Martin's Press, 1994).
14. Dalton, *Citizen Politics,* 61.
15. Ibid., 50.
16. Ibid.
17. For more on how interest groups engage in this, see Browne's chapter on "Targeting the Public," in Browne, *Groups, Interests, and U.S. Public Policy,* 84–108.
18. Mancur Olson, *The Logic of Collective Action: Public Goods and the Theory of Groups* (Cambridge: Harvard University Press, 1971). Olson is concerned with why individuals join interest groups, but much of this is applicable to NGOs as well.

19. Howard Ris, undated mass mailing letter to the author from President of Union of Concerned Scientists: Citizens and Scientists for Environmental Solutions, 2002.

20. Browne, *Groups, Interests, and Public Policy,* 209–29.

21. It should also be noted that these characteristics are important for those organizations that operate in democratic cultures, although there is potential overlap for those in non-democratic cultures as well. A comparative analysis of the similarities and differences here would make an interesting study.

22. Ramsar entered into force in 1975.

23. John Lanchbery, "Long-Term Trends in Systems for Implementation Review in International Agreements on Fauna and Flora," in *The Implementation and Effectiveness of International Environmental Commitments: Theory and Practice,* ed. David G. Victor, Kal Raustiala, and Eugene B. Skolnikoff (Cambridge: The MIT Press, 1998), 69.

24. Cyril Kormos, personal interview with Research Associate at Conservation International, Washington, D.C., October 20, 1999.

25. Estraleta Fitzhugh, personal interview with Department of Government Relations Liaison, World Wildlife Fund, Washington, D.C., July 30, 1999.

26. Scott Hajost, personal interview with Executive Director, World Conservation Union (IUNCN), Washington, D.C., October 22, 1999.

27. Sam Johnston, personal interview with Program Officer of Financial Resources and Instruments of the Convention on Biological Diversity Secretariat, Montreal, Quebec, Canada, June 30, 1999.

28. Of course, centralization does have its benefits as discussed earlier, namely the increased efficiency that it brings.

29. Center for Marine Conservation, "Center for Marine Conservation—International Coastal Cleanup." June 25, 2000, <http://www.cmc-ocean.org/cleanupbro/history.php3>.

30. Nina M. Young, "Statement of Nina M. Young, Marine Mammalogist, the Center for Marine Conservation." February 3, 2000, <http://www.environmental defense.org/pubs/Filings/96.04.30a_tuna.dolphin.htm>.

31. Estraleta Fitzhugh, personal interview with Department of Government Relations Liaison, World Wildlife Fund, Washington, D.C., July 30, 1999.

32. Save Our Environment Coalition, "Please Save America's Treasures for Future Generations!" undated advocacy letter.

33. Other members include American Rivers, League of Conservation Voters, National Audubon Society, National Parks and Conservation Association, National Wildlife Federation, Natural Resources Defense Council, the state PIRGs, and the Wilderness Society.

34. Rina Rodriguez, personal interview with International Associate, Defenders of Wildlife, Washington, D.C., July 27, 1999.

35. Defenders of Wildlife, "Number-One Nemesis," in *Defenders Annual Report 1998.* 5.

36. Shana Glickfield, personal interview with Policy Analyst Assistant, Earthjustice, Washington, D.C., July 27, 1999.

37. Mike Gordon, "Judge Won't Reverse Ruling Protecting Sea Turtles," *Planet Ark,* July 20, 2000, <http://www.planetark.org/dailynewsstory.cfm?newsid=7522>.

38. Shana Glickfield, personal interview with Policy Analyst Assistant, Earthjustice, Washington, D.C., July 27, 1999.

39. Environmental News Network Staff, "Scientists Chart a Century of Biodiversity," March 20, 2000, 4 pages, available March 24, 2000 at <http://cnn.com/2000/NATURE/03/20/biodiversity/enn/index.html>.

40. World Resources Institute, "About WRI," April 7, 1999, <http://www.wri.org/wri/wri.html>.

41. Nels Johnson, personal interview with Deputy Director, Biological Resources Program, World Resources Institute, Washington, D.C., October 20, 1999.

42. Melissa Boness, personal interview with Research Analyst at World Resources Institute headquarters, Washington, D.C., October 16, 2002.

43. Mary McClellan, personal interview with Senior Advisor for Conservation Finance, The Nature Conservancy, Arlington, Virginia, July 26, 1999.

44. Figures based upon 2003 data from IUCN's Web site: <http://www.iucn.org/members/directory.cfm>.

45. For an engaging description of this formation, see Paul Wapner's discussion of political localism and WWF in Paul Wapner, *Environmental Activism and World Civic Politics* (Albany, N.Y.: SUNY Press, 1996).

46. Ibid., 78.

47. Brooks Yeager, personal interview with Vice President, Global Threats Program, World Wildlife Fund headquarters, Washington, D.C., October 17, 2002.

48. Franklin Moore, personal interview with associate at the Environment Center, United States Agency for International Development, Washington, D.C., October 21, 1999.

49. The Sierra Club, "Amnesty International USA and the Sierra Club to Launch Historic New Campaign to Defend Environmentalists Who are Facing Human Rights Abuses," *SF Moderator,* December 6, 1999.

50. Amnesty International USA, e-mail communiqué to "The Members of AIUSA Board" from Alice Hunsberger, Director of Foundation Relations, October 14, 1998.

51. The Sierra Club, "Defending the Environmental Agenda: Take Action," *SF Moderator,* July 20, 2000.

52. Amnesty International and The Sierra Club, *Environmentalists Under Fire: Ten Urgent Cases on Human Rights Abuses,* Exec. Prod. Rory O'Connor; Writer and Prod. Stephanie Silber; Nar. Gabriel Byrne, Global Vision, 1998. 21 min.

53. Stephen Mills, personal interview with Director, International Program, Sierra Club, Washington, D.C., October 18, 2002.

54. Sierra Student Club, "About the Sierra Student Coalition," June 2, 2000, <http://www.ssc.org/aboutSSC/aboutus.html>.

55. Keith Alger, personal interview with Vice President, Conservation Strategy Department, Conservation International headquarters, Washington, D.C., October 17, 2002.

56. William Robert (Bob) Irvin, personal interview with Vice President for Marine Wildlife Conservation and General Counsel, the Ocean Conservancy, Washington, D.C., October 19, 1999.

57. For more information on GreenCOM, see their Web site at <http://www .greencom.org/>.

58. Shana Glickfield, personal interview with Policy Analyst Assistant, Earthjustice, Washington, D.C., July 27, 1999.

59. Cyril Kormos, personal interview with Research Associate at Conservation International, Washington, D.C., October 20, 1999.

60. Nicholas Lapham, personal interview with Program Officer for Environment, United Nations Foundation, Washington, D.C., July 30, 1999.

61. Numerous parties, all of whom wish to remain anonymous, expressed this concern to me.

62. Scott Hajost, personal interview with Executive Director, World Conservation Union (IUNCN), Washington, D.C., October 22, 1999.

63. "Charities You Can Trust," *SmartMoney,* December 1997 and December 1998.

64. The Trust for Public Land was the highest-rated environmental group at the close of 2002.

65. Anne Kadet, "The Agenda: Make Your Donations Count," *SmartMoney.com,* November 12, 2002.

66. American Institute of Philanthropy, "Charity Rating Guide and Watchdog Report," 32, November 2002.

67. Samuel P. Huntington, *Political Order in Changing Societies* (New Haven: Yale University Press, 1968).

68. Of interest in this study is the fact that Huntington also acknowledged the benefits of first-hand experience as a substitute for the positive effects of age of an organization. This loophole is one that participatory strategies are uniquely suited to exploit.

69. Shaiko, *Voices and Echoes for the Environment,* 111–14.

70. Ronald G. Shaiko, "The Public Interest Dilemma: Organizational Maintenance and Political Representation in the Public Interest Sector." Ph.D. dissertation, Syracuse University, 1989. Shaiko finds this true for all public interest groups, not just environmental ones.

71. Peter Uvin, "Scaling Up the Grassroots and Scaling Down the Summit: The Relations between Third World NGOs and the UN," in *NGOs, the UN, and Global Governance,* ed. Thomas G. Weiss and Leon Gordenker (Boulder: Lynne Rienner Publishers, 1996), 165.

72. Dan Seligman, personal interview with Trade and the Environment Director, the Sierra Club, Washington, D.C., October 19, 1999.

73. A similar scenario is developing within the Kyoto convention on climate

change. Despite the fact that the United States endured the worst drought of the century during the summer of 1999, nobody really felt the heat. No viable domestic political pressures exist to push a United States commitment to the levels of CO_2 reduction that are needed.

74. The Nature Conservancy, *Nature News,* April 20, 2000.

EPILOGUE: BUILDING THE NEXT ARK *(page 178–87)*

1. Qualitative and quantitative developments in transportation are another critical dimension of globalization that should not be discounted.
2. "Dust from Africa May Be Killing Caribbean Coral," *Orlando Sentinel,* September 23, 2000, A-9.
3. The old adage is that one person's terrorist is another's freedom fighter, and clearly the United States itself has engaged in its share of "freedom fighting" over the years. During the Cold War fight against the Soviet Union, it bears noting, the United States was tied to fundamentalist Muslims in Afghanistan, including the most-wanted individual, Osama bin Laden.
4. Stephen Mills, personal interview with Director, International Program, Sierra Club, Washington, D.C., October 18, 2002.
5. Theodore J. Lowi, *The End of Liberalism: The Second Republic of the United States* (New York: W.W. Norton & Company, 1979).
6. Mancur Olson, *The Rise and Decline of Nations: Economic Growth, Stagflation, and Social Rigidities* (New Haven: Yale University Press, 1982).
7. Russell J. Dalton, *Citizen Politics in Western Democracies: Public Opinion and Political Parties in the United States, Great Britain, West Germany, and France* (Chatham, N.J.: Chatham House Publishers, Inc. 1988), 109.
8. Ibid., 2.
9. Boutros Boutros-Ghali, "Foreword," in *NGOs, the UN, and Global Governance,* ed. Thomas G. Weiss and Leon Gordenker (Boulder: Lynne Rienner Publishers, 1996), 11.
10. Ibid., 7.
11. According to Genesis 9:12–17, God creates a rainbow after every rainfall as a sign of his covenant with man that a Biblical deluge will not occur again.
12. Sexism is intended here. A number of works explore the machismo attitudes prevalent in environmental destruction. One work to start with is Maria Mies and Vandana Shiva, *Ecofeminism* (London: Zed Books, 1993).
13. David W. Orr, *Ecological Literacy: Education and the Transition to a Postmodern World* (Albany: State University of New York Press, 1992).
14. Peter Herkenvath, informal personal interview with Birdlife International Associate, at the first Intersessional Meeting on Operations of the Convention on Biological Diversity, Montreal, Canada, June 29, 1999.
15. Barbara Crossette, "Private Groups to Get More Foreign Aid," *New York Times,* March 13, 1995.

16. Lorraine Elliott, *The Global Politics of the Environment* (New York: New York University Press, 1998), 145.

17. Hans Verolme, personal interview with Coordinator of Biodiversity Action Network (BIONET), Washington, D.C., headquarters, July 28, 1999.

18. Dave Foreman, *Confessions of an Eco-Warrior* (New York: Harmony Books, 1991).

19. John McCormick, "The Role of Environmental NGOs," in *The Global Environment: Institutions, Law, and Policy,* ed. Norman J. Vig and Regina S. Axelrod (Washington, D.C.: Congressional Quarterly Press, 1999), 68.

20. Stephen R. Kellert and Edward O. Wilson, *The Biophilia Hypothesis* (Washington, D.C.: Island Press, 1993). Wilson defines biophilia as the "inborn affinity human beings have for other forms of life." Edward O. Wilson, *Naturalist* (Washington, D.C.: Island Press, 1994), 360.

21. Peter H. Raven, "AIBS News: The Politics of Preserving Biodiversity," *BioScience* 40, no. 10 (November 1990): 770.

22. Ramona Rush, "Ten Tenets of Deeper Communication," *Women Transforming Communications: Global Intersections,* ed. Donna Allen, Ramona R. Rush, and Susan J. Kaufman (Thousand Oaks, California: Sage Publications, 1996).

23. John S. Dryzek, *Discursive Democracy: Politics, Policy, and Political Science* (New York: Cambridge University Press, 1994).

Bibliography

Abbey, Edward. *The Monkey Wrench Gang*. New York: Avon Books, 1975.

AIDA. "About AIDA." June 3, 2003, <http://www.aida2.org/english/index.php>.

Alger, Keith. Personal interview with Vice President, Conservation Strategy Department, Conservation International headquarters, Washington, D.C., October 17, 2002.

Amazon Watch, Project Underground, and Rainforest Action Network. "Colombia's U'wa Tribe and Supporters Celebrate Oxy's Failure to Find Oil." Press Release, July 31, 2001, <http://www.amazonwatch.org/newsroom/newsreleases01/jul31_uwa.html>.

American Institute of Philanthropy. "Charity Rating Guide and Watchdog Report," no. 32, November 2002.

American Institute of Philanthropy. "Charity Rating Guide and Watchdog Report," Spring 2000.

Astor, Michael. "Plan for Industrial Development Grows in an Ecological Paradise," *Orlando Sentinel,* June 1, 2003, G-1, 5.

Bacon, Francis. *New Atlantis*. Edited by Alfred B. Gough. Oxford: Clarendon Press, 1915.

Baldwin, David A., ed. *Neorealism and Neoliberalism: The Contemporary Debate*. New York: Columbia University Press, 1993.

Barabasi, Albert-Laszlo. *Linked: The New Science of Networks*. Cambridge, Mass.: Perseus Publishing, 2002.

Barbier, Edward B., Joanne C. Burgess, and Carl Folke. *Paradise Lost? The Ecological Economics of Biodiversity*. London: Earthscan, 1994.

Barker, Rocky. *Saving All the Parts: Reconciling Economics and the Endangered Species Act*. Washington, D.C.: Island Press, 1993.

Basinger, Julianne. "To Scientists Who Use Paying Volunteers in Fieldwork, the Benefits Outweigh the Bother." *The Chronicle of Higher Education,* June 19, 1998.

Baskin, Yvonne. *The Work of Nature: How the Diversity of Life Sustains Us*. Washington, D.C.: Island Press, 1997.

Bean, Michael. Personal interview with Director of Endangered Species and Wildlife, D.C. Congressional office, Environmental Defense, July 26, 1999.

Bech, Ulrich. Home page, *Organization & Environment: International Journal for Eco-social Research,*" January 11, 2003, <http://www.coba.usf.edu/jermier/journal.htm#Scholars>.

Beder, Sharon. *Global Spin: The Corporate Assault on Environmentalism.* Revised edition. White River Junction, Vt.: Chelsea Green Publishing Company, 2002.

Benedick, Richard. *Ozone Diplomacy: New Directions in Safeguarding the Planet.* Cambridge: Harvard University Press, 1991.

Bergst, Bret. Personal interview with Program Coordinator at World Resources Institute headquarters, Washington, D.C., October 16, 2002.

Berke, Richard L. "Sierra Club Ads in Political Races Offer a Case of 'Issue Advocacy.'" *The New York Times,* October 24, 1999, A-12.

Berry, Wendell. "The Futility of Global Thinking," *Harper's Magazine,* September 1989, 16–22. (Transcript of Commencement Address to College of the Atlantic in Bar Harbor, Maine).

Betsill, Michele M., and Elisabeth Corell. "NGO Influence in International Environmental Negotiations: A Framework for Analysis." *Global Environmental Politics* 1, no. 4 (November 2001): 65–85.

Bird, Maryann. "A Light in the Black Triangle." *Time Europe* 161, no. 17 (April 28, 2003).

Bob, Clifford. "Merchants of Morality." *Foreign Policy,* March/April 2002, 36–45.

Boness, Melissa. Personal interview with Research Analyst at World Resources Institute headquarters, Washington, D.C., October 16, 2002.

Boutros-Ghali, Boutros. "Foreword." In *NGOs, the UN, and Global Governance.* Edited by Thomas G. Weiss and Leon Gordenker. Boulder: Lynne Rienner Publishers, 1996.

Bowles, Ian A., and Cyril F. Kormos. "Environmental Reform at the World Bank: The Role of the U.S. Congress." *Virginia Journal of International Law* 35, no. 4 (Summer 1995): 777–839.

Bramble, Barbara J., and Gareth Porter. "Non-Governmental Organizations and the Making of U.S. International Environmental Policy." In *The International Politics of the Environment.* Edited by Andrew Hurrell and Benedict Kingsbury, 313–53. Oxford: Clarendon Press, 1992.

Bricker, Mindy Kay. "*Time* Names Local Scientist a 'Hero': Hydrologist Josef Krecek Uncovered Environmental Damage in North Bohemia." *The Prague Post,* May 15, 2003.

Brower, Michael, and Warren Leon. *The Consumer's Guide to Effective Environmental Choices.* New York: Three Rivers Press, 1999.

Browne, William. *Groups, Interests, and U.S. Public Policy.* Washington, D.C.: Georgetown University Press, 1998.

Bryner, Gary C. *From Promises to Performance: Achieving Global Environmental Goals.* New York: W. W. Norton & Company, 1997.

Buchanan, Mark. *Nexus: Small Worlds and the Groundbreaking Science of Networks.* New York: W.W. Norton & Company, 2002.

Buchmann, Stephen L., and Gary Paul Nabhan. *The Forgotten Pollinators.* Washington, D.C.: Island Press, 1996.

Burgiel, Stanley W., and Sheldon Cohen. *The Global Environment Facility from Rio to New Delhi: A Guide for NGOs.* Gland, Switzerland: IUCN, 1997.

The Burning Season: The Chico Mendes Story. Videocassette. Dir. John Frankenheimer, with Raul Julia. 123 min. Home Box Office, 1994.

Bush, George W. "Text of a Letter from the President to Senators Hagel, Helms, Craig, and Roberts." The White House, March 13, 2001. <http://www.whitehouse.gov/news/releases/2001/03/20010314.html>.

Caldwell, Lynton K. "Beyond Environmental Diplomacy: The Changing Institutional Structure of International Cooperation." In *International Environmental Diplomacy.* Edited by John E. Carroll. Cambridge: Cambridge University Press, 1988.

———. *International Environmental Policy: From the 20th to the 21st Century.* Durham: Duke University Press, 1996.

Camilleri, Joseph. "Rethinking Sovereignty in a Shrinking, Fragmented World." In *Contending Sovereignties: Redefining Political Community.* Edited by R. B. J. Walker and Saul H. Mendlovitz. Boulder: L. Rienner Publishers, 1990.

Cardamom's: A New Refuge for Cambodia's Wilderness. Videocassette. Exec. Prod. & Dir. Haroldo Castro. Prod. Flavia Castro. 5 min. Conservation International, 2000.

Caring for the Earth: A Strategy for Sustainable Living. Gland, Switzerland: The World Conservation Union, United Nations Program and World Wide Fund for Nature, 1991.

Carson, Rachel. *Silent Spring.* Boston: Houghton Mifflin Company, 1962.

Castro, Haroldo. Personal interview Vice President of International Communications, Conservation International headquarters, Washington, D.C., October 16, 2002.

Center for Marine Conservation. "Center for Marine Conservation—International Coastal Cleanup." June 25, 2000, <http://www.coastalcleanup.org>.

———. "National Marine Debris Monitoring Program Expands to the West Coast," 5. In *1998 Annual Report.*

Chalalan: The Story of a Dream. Videocassette. Exec. Prod. and Dir. Haroldo Castro. Prod. Flavia Castro. Nar. Luis Manrique. 5 min. Conservation International, 2000.

Chambers, Nina M., and Sam H. Ham. "Strengthening Regional Planning through Community Education." In *Conservation of Biodiversity and the New Regional Planning.* Edited by Richard E. Saunier and Richard A. Meganck, 75–92. Organization of American States and The World Conservation Union, 1995.

"Charities You Can Trust," *SmartMoney,* December 1998.

Clifton, Carr, Tom Turner, and Stewart L. Udall. *Wild by Law: The Sierra Club Legal Defense Fund and the Places It Has Saved.* San Francisco: Sierra Club Books, 1990.

Cohen, Michael P. *The History of the Sierra Club: 1892–1970.* San Francisco: Sierra Club Books, 1988.

Coleman, James. *Foundations of Social Theory.* Cambridge: Harvard University, 1990.

Commission on Global Governance. *Our Global Neighbourhood.* New York: Oxford University Press, 1995.

Conca, Ken, and Geoffrey D. Dabelko, *Green Planet Blues: Environmental Politics from Stockholm to Kyoto.* Boulder: Westview Press, 1998.

Conservation International. "Creating Solutions for the Twenty-First Century: 1998 Annual Report."

———. "Film Capturing Guyana's Natural Heritage a Finalist at Jackson Hole." Press release, September 23, 1999.

———. "Monitoring and Evaluation Program: Performance and Management." June 25, 2000, <http://www.conservation.org/WEB/ABOUTCI/Monitor.htm>.

———. "Promoting Conservation through International Assistance." June 25, 2000, <http://www.conservation.org/WEB/FIELDACT/C-C_PROG.policy.intlasst.htm> [URL no longer available].

"Convention on Biological Diversity." January 18, 1999, <http://www.unep.ch/bio/conve.html>.

Corell, Elisabeth, and Michele M. Betsill. "A Comparative Look at NGO Influence in International Environmental Negotiations: Desertification and Climate Change." *Global Environmental Politics* 1, no. 4 (November 2001): 86–107.

Cosgrove, Dennis. "Contested Global Visions: One-World, Whole Earth, and the Apollo Space Program." *Annals of the Association of American Geographers* 84, no. 2: 270–94.

Crossette, Barbara. "Private Groups to Get More Foreign Aid." *New York Times,* March 13, 1995.

Daily, Gretchen, ed. *Nature's Services: Societal Dependence on Natural Ecosystems.* Washington, D.C.: Island Press, 1997.

Dalton, Russell J. *Citizen Politics in Western Democracies: Public Opinion and Political Parties in the United States, Great Britain, West Germany, and France.* Chatham, N.J.: Chatham House Publishers, Inc. 1988.

Daly, Herman E. *Beyond Growth: The Economics of Sustainable Development.* Boston: Beacon Press, 1996.

Daly, Herman, and John Cobb, Jr. *For the Common Good: Redirecting the Economy Toward Community, the Environment, and a Sustainable Future.* Boston: Beacon Press, 1989.

Day, David. *The Whale War.* San Francisco: Sierra Club Books, 1987.

Defenders of Wildlife. *Defenders Annual Report 1998.*

————. "DENlines Issue #13 (Extended Earth Day Issue)." On-line posting. Discussion listserv. April 21, 2000, <http://www.denaction.org/>.

DeLuca, Kevin Michael. *Image Politics: The New Rhetoric of Environmental Activism*. New York: The Guilford Press, 1999.

Dennis, Everette E. "In Context: Environmentalism in the System of News." *Media and the Environment*. Edited by Craig L. LaMay and Everette E. Dennis. Washington, D.C.: Island Press, 1991.

Descartes, René. *Discourse on the Method*. Edited by David Weissman. New Haven: Yale University Press, 1996.

Dobson, Andrew P. *Conservation and Biodiversity*. New York: Scientific American Library, 2000.

A Dream for Guyana's National Heritage. Videocassette. Exec. Prod. & Dir. Haroldo Castro. Prod. Flavia Castro. Host Kojo Nnamdi. 15 min. Conservation International, 1998–1999.

Dryzek, John S. *Discursive Democracy: Politics, Policy, and Political Science*. New York: Cambridge University Press, 1994.

"Dust from Africa May Be Killing Caribbean Coral." *Orlando Sentinel*, September 23, 2000, A-9.

Earthjustice Legal Defense Fund. "About Earthjustice." [Oakland, Calif.]: n.p., May 30, 1999, <http://www.earthjustice.org/about/index.html>.

————. "Annual Report 1998." [Oakland, Calif.]: n.p., 1999. 12–13.

————. "Docket 2002." [Oakland, Calif.]: n.p., 2001.

————. "Docket 2000." [Oakland, Calif.]: n.p., 1999.

————. "Docket 1999." [Oakland, Calif.]: n.p., 1998.

Earthwatch Institute. "2002–2003 Research and Exploration," 21, no. 3: 54.

————. "The Center for Field Research at Earthwatch Institute Applications Guidelines Summary."

————. *Earthwatch Institute 1999 Research and Exploration*. Watertown, Mass.: Earthwatch Expedition, Inc. 1999.

————. "Education Programs at Earthwatch," June 25, 2000, <http://www.earthwatch.org/aboutew/education.html>.

————. "How Earthwatch Works." In *Earthwatch Institute 1999 Research and Exploration*. Watertown, Mass.: Earthwatch Expedition, Inc. 1999.

————. "Global Classroom." August 26, 1999, <http://www.earthwatch.org/ed/olr/biodiversity.html>.

Economy, Elizabeth, and Miranda A. Schreurs. "Domestic and International Linkages in Environmental Politics." In *The Internationalization of Environmental Protection*. Edited by Miranda A. Schreurs and Elizabeth C. Economy. New York: Cambridge University Press, 1997.

Ehrlich, Paul R. *The Machinery of Nature: The Living World Around Us—And How It Works*. New York: Simon and Schuster, 1986.

Eldredge, Niles. *Life in the Balance: Humanity and the Biodiversity Crisis*. Princeton: Princton University Press, 1998.

Elkins, David J., and Richard E.B. Simeon. "A Cause in Search of its Effect, or What Does Political Culture Explain." *Comparative Politics.* 11 (January 1979): 127–46.

Elliott, Jennifer. *An Introduction to Sustainable Development.* New York: Routledge, 1999.

Elliott, Lorraine. *The Global Politics of the Environment.* New York: New York University Press, 1998.

Environmental Defense Fund. "EDF's International Program at Work." [Eight-page glossy pamphlet.] n.p., n.d.

———. "Action Alert." On-line posting. Discussion listserv, June 1, 2000, <http://www.environmentaldefense.org/actioncenter.cfm>.

———. "Environmental Defense Action Network 1999 Wrap Up." On-line posting. Discussion Listserv. February 4, 2000, EDF@actionnetwork.org.

———. "Who We Are, What We Do, How You Can Help: A Guide for Members and Friends." Undated pamphlet.

Environmental News Network Staff. "Scientists Chart a Century of Biodiversity." March 24, 2000, <http://cnn.com/2000/NATURE/03/20/biodiversity/enn/index.html>.

Environmentalists Under Fire: Ten Urgent Cases on Human Rights Abuses. Exec. Prod. Rory O'Connor. Writer and Prod. Stephanie Silber. Nar. Gabriel Byrne. Amnesty International and The Sierra Club. 21 min. Global Vision, 1998.

Falk, Richard. *This Endangered Planet: Prospects and Proposals for Human Survival.* New York: Vintage Books, 1971.

Fitzhugh, Estraleta. Personal interview with Department of Government Relations Liaison, World Wildlife Fund, Washington, D.C., July 30, 1999.

Foreman, Dave. *Confessions of an Eco-Warrior.* New York: Harmony Books, 1991.

Foster-Turley, Pat. *Making Biodiversity Happen: The Role of Environmental Education and Communication.* Washington, D.C.: GreenCOM, 1996.

Fowler, Michael Ross, and Julie Marie Bunck. *Law, Power, and the Sovereign State: The Evolution and Application of the Concept of Sovereignty.* University Park: Pennsylvania State University, 1995.

Fox, Jonathan A., and L. David Brown, eds. *The Struggle for Accountability: The World Bank, NGOs, and Grassroots Movements.* Cambridge: MIT Press, 1998.

Frank, Andre Gunder. *Capitalism and Underdevelopment in Latin America: Historical Studies of Chile and Brazil.* New York: Monthly Review Press, 1967.

Friedman, Milton. *Free to Choose: A Personal Statement.* New York: Harcourt Brace Jovanovich, 1980.

Fuller, Kathryn S. "President's Message: Thinking and Acting Globally and Locally," *Focus* 22, no. 1 (January/February 2000) 2.

"The GEF in the 21st Century: A Vision for Strengthening the Global Environment Facility." December 20, 1999, <www.igc.org/bionet/gef21.html> [URL no longer available].

Gelbspan, Ross. *The Heat is On: The Climate Crisis, The Cover-Up, The Prescription*. Reading, Mass.: Perseus Books, 1998.

Gibbs, Lois Marie. *Love Canal: The Story Continues*. Stony Creek, Conn.: New Society Publishers, 1998.

Gifford, Bill. "The Greening of the Golden Arches." In *The Rolling Stone Environmental Reader*, 216–23. Washington, D.C.: Island Press, 1992.

Gilpin, Robert. "Three Models of the Future." In *Transnational Corporations and World Order*. Edited by George Modelski. San Francisco: W.H. Freeman, 1979.

———. *War and Change in World Politics*. Cambridge: University Press, 1981.

Glacken, Charles J. *Traces on the Rhodian Shore: Nature and Culture in Western Thought from Ancient Times to the End of the Eighteenth Century*. Berkeley: University of California Press, 1967.

Gladwell, Malcolm. *The Tipping Point: How Little Things Can Make a Big Difference*. Boston: Back Bay Books, 2002.

Glesne, Corrine, and Alan Peshkin. *Becoming Qualitative Researchers: An Introduction*. White Plains, N.Y.: Longman, 1992.

Glickfield, Shana. Personal interview with Policy Analyst Assistant, Earthjustice Legal Defense Fund, Washington, D.C., July 27, 1999.

Goldfarb, Theodore, ed. *Notable Selections in Environmental Studies*. 2nd ed. Guildford, Conn.: Dushkin/McGraw-Hill, 2000.

Gonzalez, George A. *Corporate Power and the Environment: The Political Economy of U.S. Environmental Policy*. Lanham, Md.: Rowman & Littlefield, 2001.

Gordenker, Leon, and Thomas G. Weiss. "Pluralizing Global Governance: Analytical Approaches and Dimensions." In *NGOs, the UN, and Global Governance*. Edited by Thomas G. Weiss and Leon Gordenker. Boulder: Lynne Rienner Publishers, 1996.

Gordon, Mike. "Judge Won't Reverse Ruling Protecting Sea Turtles." *Planet Ark*, July 20, 2000, <http://www.planetark.org/dailynewsstory.cfm?newsid=7522>.

Gore, Al. *Earth in the Balance: Ecology and the Human Spirit*. Boston: Houghton Mifflin Company, 1992.

Gould, Stephen Jay. *Eight Little Piggies: Reflections in Natural History*. New York: W.W. Norton & Company, 1993.

———. *Full House: The Spread of Excellence from Plato to Darwin*. New York: Three Rivers Press, 1996.

———. *Wonderful Life: The Burgess Shale and the Nature of History*. New York: W.W. Norton & Company, 1989.

Gourevitch, Peter. *Politics in Hard Times: Comparative Responses to International Economic Crises*. Ithaca: Cornell University Press, 1986.

———. "The Second Image Reversed: The International Sources of Domestic Politics." *International Organization* 32, no. 4 (Autumn 1978): 881–911.

Gray, Charles. "Corporate Goliaths: Sizing Up Corporations and Governments." *Multinational Monitor*, June 1999, 26–27.

Green Earth Organization et al. "Statement of the 14th Global Biodiversity Forum," June 18–20, 1999, Montreal, Canada.

Greico, Joseph. "Anarchy and the Limits of Cooperation: A Realist Critique of the Newest Liberal Institutionalism." *International Organization.* 42 (August 1988): 485–507.

Grove, Noel. "Quietly Conserving Nature." *National Geographic* 174, no. 6 (December 1988): 818–844.

Grumbine, R. Edward, ed. *Environmental Policy and Biodiverstiy.* Washington, D.C.: Island Press, 1994.

———. *Ghost Bears: Exploring the Biodiversity Crisis.* Washington, D.C.: Island Press, 1992.

Guggenheim, David. Personal interview with Vice President for Conservation Policy, The Ocean Conservancy headquarters, Washington, D.C., October 17, 2002.

Guruswamy, Lakshman D., and Jeffrey A. McNeely, ed. *Protection of Global Biodiversity: Converging Strategies.* Durham: Duke University Press, 1998.

Haas, Peter. *Saving the Mediterranean: The Politics of International Environmental Cooperation.* New York: Columbia University Press, 1990.

Haas, Peter M., Robert O. Keohane, and Marc A. Levy. *Institutions for the Earth: Sources of Effective International Environmental Protection.* Cambridge: The MIT Press, 1994.

Hagan, Elizabeth. "Three Babes and a Bus Named 'Bella:' The Sierra Student Coalition Rainforest Education Bus Mania Swept the East Coast this May!" *Generation E: The Voice of the Sierra Student Coalition,* Summer 1999, 4.

Hajost, Scott. Personal interview with Executive Director, World Conservation Union (IUNCN), Washington, D.C., October 22, 1999.

Harris, Lis. *Tilting at Mills: Green Dreams, Dirty Dealings, and the Corporate Squeeze.* New York: Houghton Mifflin, 2003.

Hawken, Paul. *The Ecology of Commerce: A Declaration of Sustainability.* New York: HarperCollins Publishers, 1993.

Hecht, Susanna, and Alexander Cockburn. *The Fate of the Forest: Developers, Destoyers and Defenders of the Amazon.* London: Verson, 1989.

Hempel, Lamont C. *Environmental Governance: The Global Challenge.* Washington, D.C.: Island Press, 1996.

Henderson, Hazel. *The Politics of the Solar Age: Alternatives to Economics.* Indianapolis: Knowledge Systems, 1988.

Herkenvath, Peter. Informal interview with Birdlife International representative at the first Intersessional Meeting on Operations of the Convention on Biological Diversity, Montreal, Quebec, Canada, June 29, 1999.

Hightower, Jim. "Get the Hogs Out of the Creek!" *Earth Island Journal.* 11, no. 1 (1995): 32.

Hochstetler, Kathryn. "After the Boomerang: Environmental Movements and Politics in the La Plata River Basin." *Global Environmental Politics.* 2, no. 4 (November 2002): 35–57.

Homer-Dixon, Thomas, and Jessica Blitt, eds. *Ecoviolence: Links Among Environment, Population, and Security.* Lanham, Md.: Rowman & Littlefield Publishers, Inc., 1998.

Honey, Martha S. "Treading Lightly? Ecotourism's Impact on the Environment," *Environment* 41, no. 5 (June 1999): 4–9, 28–33.

Horta, Korinna. Personal interview with Senior Environmental Economist, Environmental Defense, Washington, D.C., October 18, 2002.

Horta, Korinna. "Rhetoric and Reality: Human Rights and the World Bank." *Harvard Human Rights Journal.* 15 (Spring 2002): 227–43.

Hotspots: Protecting Earth's Most Endangered Treasures. Videocassette. Exec. Prod. & Dir. Haroldo Castro. Prod. Flavia Castro. Host Dr. Russell Mittermeier. Nar. Paul Anthony. 12 min. Conservation International, 1999–2000.

Hubert, Don. "Inferring Influence: Gauging the Impact of NGOs." In *Toward Understanding Global Governance: The International Law and International Relations Toolbox.* Edited by Charlotte Ku and Thomas G. Weiss, 27–51. Providence, R.I.: ACUNS/Reports and Papers 1998, No. 2.

Hughes, Barry B. *Continuity and Change in World Politics: The Clash of Perspectives.* Englewood Cliffs, N.J.: Prentice-Hall, 1991.

Hunsberger, Alice. Director of Foundation Relations, Amnesty International USA. Email communiqué to "The Members of AIUSA Board," October 14, 1998.

Huntington, Samuel P. *Political Order in Changing Societies.* New Haven: Yale University Press, 1968.

Hurrell, Andrew. "International Society and the Study of Regimes: A Reflective Approach." In *Regime Theory and International Relations.* Edited by Volker Rittberger. Oxford: Clarendon Press, 1995.

Irvin, Robert. Personal interview with Vice President Marine Wildlife Conservation and General Counsel, the Center for Marine Conservation, Washington, D.C., October 19, 1999.

Jasanoff, Sheila. Informal conversation at the annual Association for Politics and the Life Sciences convention, Boston, Mass., September 1998.

———. "NGOs and the Environment: From Knowledge to Action." *Third World Quarterly* 18, no. 3 (December 1997): 579–94.

Johnson, Nels. Personal interview with Deputy Director, Biological Resources Program, World Resources Institute, Washington, D.C., October 20, 1999.

Johnston, Sam. Personal interview with Program Officer of Financial Resources and Instruments. Convention on Biological Diversity Secretariat, Montreal, Quebec, June 30, 1999.

Jolly, Alison. "The Madagascar Challenge: Human Needs and Fragile Ecosystems." In *Environment and the Poor: Development Strategies for a Common Agenda.* Edited by H. Jeffrey Leonard. New Brunswick, N.J.: Transaction Books, 1989.

Jones, Charles O. "Doing Before Knowing: Concept Development in Political Research." In *Theory-Building and Data Analysis in the Social Sciences.* Edited by

Herbert B. Asher, Herbert F. Weisberg, John H. Kessel, and W. Phillips Shively. Knoxville: The University of Tennessee Press, 1984.

Kadet, Anne. "The Agenda: Make Your Donations Count." *SmartMoney.com*, November 12, 2002.

Kahn, Herman. *The Next Two Hundred Years: A Scenario for America and the World*. New York: Morrow, 1976.

Kaul, Inge, Isabelle Grunberg, and Marc A. Stern, eds. *Global Public Goods: International Cooperation in the Twenty-First Century*. New York: Oxford University Press, 1999.

Kegley, Charles W., and Eugene R. Wittkopf. *World Politics: Trend and Transformation*. 7th ed. Boston: Bedford, 1999.

Kellert, Stephen R. *The Value of Life: Biological Diversity and Human Society*. Washington, D.C.: Island Press, 1996.

Kellert, Stephen R. and Edward O. Wilson, eds. *The Biophilia Hypothesis*. Washington, D.C.: Island Press, 1993.

Keohane, Robert O. *Neorealism and Its Critics*. New York: Columbia University Press, 1986.

Keohane, Robert O., and Joseph S. Nye. *Power and Interdependence: World Politics in Transition*. Boston: Little, Brown and Company, Inc., 1977.

Klare, Michael. *Resource Wars: The New Landscape of Conflict*. New York: Metropolitan Books, 2001.

Kormos, Cyril. Personal interview with Research Associate at Conservation International headquarters, Washington, D.C., October 20, 1999.

Krasner, Stephen. "Westphalia and All That." In *Ideas and Foreign Policy: Beliefs, Institutions and Political Change*. Edited by Judith Goldstein and Robert Keohane. Ithaca: Cornell University Press, 1993.

Krecek, Joseph. Associate Professor Lecturer Institute of Natural Resources, European University Department of Hydrobiology. Letter to Mountain Waters of Bohemia prospective participants. Prague, Czech Republic, November 23, 2001. Earthwatch Institute's "Meet the Scientists." <http://www.earthwatch.org/expeditions/krecek/meetthescientists.html#staff>.

Krueger, Richard A. *Developing Questions for Focus Groups: Focus Group Kit 3*. London: Sage Publications, 1998.

Kunich, John Charles. *Ark of the Broken Covenant: Protecting the World's Biodiversity Hotspots*. Westport, Conn.: Praeger, 2003.

Lanchbery, John. "Long-Term Trends in Systems for Implementation Review in International Agreements on Fauna and Flora." In *The Implementation and Effectiveness of International Environmental Commitments: Theory and Practice*. Edited by David G. Victor, Kal Raustiala, and Eugene B. Skolnikoff, 57–87. Cambridge: The MIT Press, 1998.

Lapham, Nicholas. Personal interview with Program Officer for Environment, United Nations Foundation, Washington, D.C., July 30, 1999.

Leopold, Aldo. *A Sand County Almanac: And Sketches Here and There*. London: Oxford University Press, 1949.

Lind, Don. "The Earth Home We See from Space." *National Forum* 75, no. 1 (Winter 1995).

Lipschutz, Ronnie D., and Ken Conca, eds. *The State and Social Power in Global Environmental Politics.* New York: Columbia University Press, 1993.

Litfin, Karen. "Eco-regimes: Playing Tug of War with the Nation-State." In *The State and Social Power in Global Environmental Politics.* Edited by Ronnie D. Lipschutz and Ken Conca. New York: Columbia University Press, 1993.

———. *The Greening of Sovereignty in World Politics.* Cambridge: The MIT Press, 1998.

———. "Sovereignty in World Ecopolitics." *Mershon International Studies Review* 41, no. 2 (November 1997).

Lovejoy, Thomas. "Aid Debtor Nation's Ecology," *New York Times,* October 4, 1984.

Lovelock, James E. *Gaia: A New Look at Life on Earth.* Oxford: Oxford University Press, 1979.

Lovins, Amory B. "Cost-Risk-Benefit Assessments in Energy Policy." *The George Washington Law Review* 45, no. 5 (August 1977): 911–42.

———. *Soft Energy Paths: Toward a Durable Peace.* Cambridge, Mass.: Ballinger Pub. Co., 1977.

Lowi, Theodore J. *The End of Liberalism: The Second Republic of the United States.* New York: W.W. Norton & Company, 1979.

Luke, Tim. "The Nature Conservancy or the Nature Cemetery: Buying and Selling 'Perpetual Care' as Environmental Resistance." *Capitalism Nature and Socialism: A Journal of Socialist Ecology* 6, no. 2 (June 1995): 1–20.

Magruder, Blue. Personal interview with Public Relations Director at Earthwatch Institute headquarters, Watertown, Mass., July 8, 1999.

Mann, Charles, and Mark Plummer. *Noah's Choice: The Future of Endangered Species.* New York: Alfred A. Knopf, 1995.

Marsh, George Perkins. *Man and Nature.* Edited by David Lowenthal. Cambridge: Belknap Press of Harvard University Press, 1965.

Martens, Jens. "NGOs in the UN System: The Participation of Non-Governmental Organizations in Environment and Development Institutions of the United Nations." Projectstelle UNCED, DNR/BUND, Bonn, September 1992.

Marx, Leo. "American Institutions and Ecological Ideals." *Science* 170 (November 27, 1970): 945–52.

———. "Technology: The Emergence of a Hazardous Concept." *Social Research* 64, no. 3 (Fall 1997): 965–88.

Mast, Roderic. Telephone interview with Vice President, CI-Sojourns, Conservation International, June 5, 2003.

Mathews, Jessica T. "Power Shift: The Rise of Global Civil Society." *Foreign Affairs* 76 (January/February 1997): 50– 66.

McClellan, Mary. Personal interview with Senior Advisor for Conservation Finance, The Nature Conservancy, Arlington, Va., July 26, 1999.

McCormick, John. "The Role of Environmental NGOs." *In The Global Environ-*

ment: Institutions, Law, and Policy. Edited by Norman J. Vig and Regina S. Axelrod. Washington, D.C.: Congressional Quarterly Press, 1999.

McKee, Jeffrey K. *Sparing Nature: The Conflict between Human Population Growth and Earth's Biodiversity.* Piscataway, N.J.: Rutgers University Press, 2003.

McKibben, Bill. *The End of Nature.* New York: Random House, 1989.

McNeely, Jeffrey A., Kenton R. Miller, Walter V. Reid, Russell A. Mittermeier and Timothy Werner. "Strategies for Conserving Biodiversity." *Environment* 32, no. 3 (April 1990): 18.

McNeill, J. R. *Something New Under the Sun: An Environmental History of the Twentieth-Century World.* New York: W.W. Norton & Company, 2000.

McPhee, John. *Encounters with the Archdruid.* New York: Farrar, Straus and Giroux, 1971.

Meacham, Cory J. *How the Tiger Lost Its Stripes: An Exploration into the Endangerment of a Species.* New York: Harcourt Brace & Company, 1997.

Meadows, Donella. *The Limits to Growth: A Report for the Club of Rome's Project on the Predicament of Mankind.* New York: Universe Books, 1972.

Merchant, Carolyn. *Radical Ecology: The Search for a Livable World.* New York: Routledge, 1992

Mies, Maria, and Vandana Shiva. *Ecofeminism.* London: Zed Books, 1993.

Miller, Marian A. L. *The Third World in Global Environmental Politics.* Boulder: Lynne Rienner, 1995.

Mills, C. Wright. *The Power Elite.* New York: Oxford University Press, 1956.

Mills, Stephen. Personal interview with Director, International Program, Sierra Club, Washington, D.C., October 18, 2002.

Mingst, Karen. "Implementing International Environmental Treaties: The Role of NGOs." Paper presented at the annual meeting of the International Studies Association, Mexico, 1993.

———. "Uncovering the Missing Links: Linkage Actors and their Strategies in Foreign Policy Analysis." In *Foreign Policy Analysis: Continuity and Change in Its Second Generation.* Edited by Laura Neack, Jeanne A. K. Hey, and Patrick J. Haney. Englewood Cliffs, N.J.: Prentice Hall, 1995.

Moore, Franklin. Personal interview with associate at the Environment Center, United States Agency for International Development, Washington, D.C., October 21, 1999.

Monroe, Martha C., ed. *What Works: A Guide to Environmental Education and Communication Projects for Practitioners and Donors.* Washington, D.C.: Academy for Educational Development, 1999.

Morell, Virginia. "In Search of Solutions." *National Geographic* 195, no. 2 (February 1999): 72–87.

———. "Restoring Madagascar," *National Geographic* 195, no. 2 (February 1999): 60–71.

———. "The Sixth Extinction." *National Geographic* 195, no. 2 (February 1999): 42–59.

————. "The Variety of Life." *National Geographic* 195, no. 2 (February 1999): 6–31.

————. "Wilderness Headcount." *National Geographic* 195, no. 2 (February 1999): 32–41.

Morgenthau, Hans. *Politics among Nations: The Struggle for Power and Peace.* New York: Alfred A. Knopf, 1948.

Muffett, Carroll. Personal interview with International Programs Director, Defenders of Wildlife headquarters, Washington, D.C., October 17, 2002.

Muttlingyam, Sanjayan. Personal interview with Lead Scientist, The Nature Conservancy headquarters, Arlington, Va., October 16, 2002.

Nash, Roderick, ed. *Grand Canyon of the Living Colorado.* New York: Sierra Club/ Ballantine Books, 1970.

The Nature Conservancy. "International Trips 1999." n.p., n.d.

————. *Nature News,* April 20, 2000.

Nef, John U. "An Early Energy Crisis and Its Consequences." *Scientific American* 237 (November 1977): 140–51.

Nickens, T. Edward. "High Adventure, Low Impact." *Cooking Light,* July/August 1998, 80–85.

Nye, Joseph S. "Comparing Common Markets: A Revised Neo-Functionalist Model." In *International Organization: A Reader.* Edited by Friedrich Kratochwil and Edward O. Mansfield. New York: HarperCollins, 1994.

O'Brien, Anne. "Purchasing Power: Why We *Still* Buy Land." *Nature Conservancy,* November/December 1999, 12–17.

The Ocean Conservancy. "National Marine Debris Monitoring Program Expands to the West Coast." *1998 Annual Report,* (1999): 5.

Ohmae, Kenichi. *The Borderless World: Power and Strategy in the Interlinked Economy.* New York: HarperBusiness, 1990.

Olsen, Jennifer, and Thomas Princen. "Hazardous Waste Trade, North and South: The Case of Italy and Koko, Nigeria." School of Foreign Service, Georgetown University, 1994.

Olson, Mancur. *The Logic of Collective Action: Public Goods and the Theory of Groups.* Cambridge: Harvard University Press, 1971.

————. *The Rise and Decline of Nations: Economic Growth, Stagflation, and Social Rigidities.* New Haven: Yale University Press, 1982.

Ophuls, William. *Ecology and the Politics of Scarcity.* San Francisco: W.H. Freeman and Company, 1977.

Orr, David W. *Earth in Mind: On Education, Environment, and the Human Prospect.* Washington, D.C.: Island Press, 1994.

————. *Ecological Literacy: Education and the Transition to a Postmodern World.* Albany: State University of New York Press, 1992.

Ottaway, David B., and Joe Stephens. "Nonprofit Land Bank Amasses Billions: Charity Builds Assets on Corporate Partnerships." *The Washington Post,* May 4, 2003, A-1.

Paarlberg, Robert L. "A Domestic Dispute: Clinton, Congress and International Environmental Policy." *Environment* 38, no. 8 (October 1996): 18.

Panjabi, Ranee K. L. *The Earth Summit at Rio: Politics, Economics, and the Environment.* Boston: Northeastern University Press, 1997.

Patterson, Alan. "Debt for Nature Swaps: And the Need for Alternatives." *Environment* 32, no. 10 (December 1990): 5–13, 31–32.

Paul, Seema. Personal interview with Senior Program Officer, Biodiversity, at United Nations Foundation headquarters, Washington, D.C., October 18, 2002.

Perlman, Dan L., and Glenn Adelson. *Biodiversity: Exploring Values and Priorities in Conservation.* Cambridge, Mass.: Blackwell Science, 1997.

Phillips, Kathryn. *Tracking the Vanishing Frogs: An Ecological Mystery.* New York: St. Martin's Press, 1994.

Porter, Gareth, and Janet Welsh Brown. *Global Environmental Politics.* Boulder: Westview Press, 1996.

Posey, Darrell A. "Protecting Indigenous People's Rights to Biodiversity: People, Property and Bio-prospecting." *Environment* 38, no. 8 (October 1996).

Poupon, Christine. "NGOs Aim to Influence UN on Environment and Development." *CERES* 24, no. 2 (March–April 1992).

Prakash, Reshma. Personal interview at *The Earth Times* headquarters, New York City, August 3, 1999.

Preston, Richard. *The Hot Zone.* New York: Random House, 1994.

Princen, Thomas, and Mathias Finger. *Environmental NGOs in World Politics: Linking the Local and the Global.* London: Routledge, 1994.

Putnam, Robert. "Diplomacy and Domestic Politics: The Logic of Two Level Games." *International Organization* 42, no. 3 (Summer 1988): 427–60.

———. *Making Democracy Work: Civic Traditions in Modern Italy.* Princeton, N.J.: Princeton University Press, 1993.

Quammen, David. *The Song of the Dodo: Island Biogeography in an Age of Extinction.* New York: Simon & Schuster, 1996.

Rakotovao, Lala H., R. Rakotoariseheno, and Chantale Andrianarivo. "Conservation in Action: Assessing the Behaviour of National and International Researchers Working in Madagascar." In *Protecting Biological Diversity: Roles and Responsibilities.* Edited by Catherine Potvin, Margaret Kraenzel, and Gilles Seutin. Montreal: McGill-Queen's University Press, 2001.

Raustiala, Kal. "States, NGOs, and International Environmental Institutions," *International Studies Quarterly* 41, no. 4 (December 1997): 719–40.

Raustiala, Kal, and David G. Victor. "Biodiversity Since Rio: The Future of the Convention on Biological Diversity." *Environment* 38, no. 4 (May 1996).

Raven, Peter H. "AIBS News: The Politics of Preserving Biodiversity." *BioScience* 40, no. 10 (November 1990).

Reaka-Kudla, Marjorie L., Don E. Wilson, and Edward O. Wilson, eds. *Biodiversity II: Understanding and Protecting Our Biological Resources.* Washington, D.C.: Joseph Henry Press, 1997.

Redesigning the Landscape: Cerrado-Pantanal Conservation Corridor. Video-cassette. Exec. Prod. & Dir. Haroldo Castro. Prod. Flavia Castro. Host Osmar Bastos. 11 min. Conservation International, 2002.

Rich, Bruce. *Mortgaging the Earth: The World Bank, Environmental Impoverishment and the Crisis of Development.* Boston: Beacon Press, 1994.

Rich , Bruce. Personal interview Program Manager, International Program, Environmental Defense, Capitol Hill office, Washington, D.C., October 22, 1999.

Ris, Howard. Undated mass mailing letter to the author from President of Union of Concerned Scientists: Citizens and Scientists for Environmental Solutions, 2002.

Robbins, Michel W. "Biodiversity and Strange Bedfellows." Editorial. *Audubon* 97, no. 1 (January–February 1995): 4.

Rodriguez, Rina. Personal interview with International Associate, Defenders of Wildlife headquarters, Washington, D.C., July 27, 1999.

Rogowski, Ronald. *Commerce and Coalitions: How Trade Affects Domestic Politics Alignments.* Princeton: Princeton University Press, 1989.

Rohrschneider, Robert, and Russell J. Dalton. "A Global Network? Transnational Cooperation among Environmental Groups." *The Journal of Politics* 64, no. 2 (May 2002): 510–533.

Rosenau, James N. "Governance, Order and Change in World Politics." In *Governance without Government: Order and Change in World Politics.* Edited by James N. Rosenau and E. O. Czempiel. Cambridge: Cambridge University Press, 1994.

———. *Linkage Politics: Essays on the Convergence of National and International Systems.* New York: Free Press, 1969.

Rush, Ramona. "Ten Tenets of Deeper Communication." In *Women Transforming Communications: Global Intersections.* Edited by Donna Allen, Ramona R. Rush, and Susan J. Kaufman. Thousand Oaks, Calif.: Sage Publications, 1996.

Save Our Environment Coalition. "Please Save America's Treasures for Future Generations!" undated advocacy letter.

Sawhill, John C. "The Nature Conservancy." *Environment* 38, no. 5 (June 1996): 43–44.

Say No to Bushmeat: Stop Killing Wild Animals. Videocassette. Exec. Prod. & Dir. Haroldo Castro. Prod. and Ed. Flavia Castro. Nar. Akwei Thompson. 9 min. Conservation International, 2002.

Scenic Hudson Preservation Conference v. Federal Power Commission, 453 F2nd 463.

Schlickeisen, Rodger. "The Internet Revolution." *Defenders,* Spring 2000, <http://www.defenders.org/magazinenew/rs/rssp00.html>.

Schnaiberg, Allan, and Kenneth Alan Gould. *Environment and Society: The Enduring Conflict.* New York: St. Martin's Press, 1994.

Schumacher, E. F. *Small is Beautiful: Economics as if People Mattered.* London: Blond and Briggs, 1973. (Reprint, New York: HarperPerennial, 1989).

Segal, David. "A Nation of Lobbyists." *The Washington Post National Weekly Edition* 12, no. 37 (1995): 11.

Seidl, John M. (Mick). Telephone interview with Chief Program Officer, Environment, Gordon and Betty Moore Foundation, San Francisco, Calif., June 4, 2003.

Seidman, I. E. *Interviewing as Qualitative Research: A Guide for Researchers in Education and the Social Sciences.* New York: Teachers College Press, 1991.

Seligman, Dan. Personal interview at the Sierra Club legislative headquarters, Washington, D.C., October 19, 1999.

Shaiko, Ronald G. "The Public Interest Dilemma: Organizational Maintenance and Political Representation in the Public Interest Sector." Ph.D. dissertation, Syracuse University, 1989.

———. *Voices and Echoes for the Environment: Public Interest Representation in the 1990s and Beyond.* New York: Columbia University Press, 1999.

Sheehan, James, and Paul Georgia. "Feeding the Green Money Tree." *The Washington Times,* July 29, 1999, A-20.

Shiva, Vandana. "The Greening of Global Reach." In *Global Ecology: A New Arena of Conflict.* Edited by Wolfgang Sachs, 145–56. London: Zed Books, 1993.

Siegel, Tatiana. "Biodiversity and You." *The Nation,* March 17, 2003, 32–36.

Sierra Club. "Amnesty International USA and the Sierra Club to Launch Historic New Campaign to Defend Environmentalists Who are Facing Human Rights Abuses." *SF Moderator.* On-line posting, December 6, 1999.

———. "Defending the Environmental Agenda: Take Action." *SF Moderator.* On-line posting, July 20, 2000.

———. *Grassroots Organizing Training Manual.* San Francisco: Sierra Club, 1999.

Sierra Legal Defence Fund. "About Us." June 3, 2003, <http://www.sierralegal.org/aboutsierralegal.html>.

Sierra Student Club. "About the Sierra Student Coalition." June 2, 2000, <http://www.ssc.org/aboutSSC/aboutus.html>.

Simon, Julian. *The Ultimate Resource.* Princeton, N.J.: Princeton University Press, 1981.

Slaughter, Anne-Marie. "The Real New World Order." *Foreign Affairs* 76 (September/October 1997): 183–97.

Solbrig, Otto T. "The Origin and Function of Biodiversity." *Environment* 33, no. 5 (June 1991).

Soroos, Marvin S. *Beyond Sovereignty: The Challenge of Global Policy.* Columbia: University of South Carolina Press, 1986.

South and Meso American Indian Rights Center (SAIIC). "U'wa of Colombia Reject All New Oil Exploration: New Report Details Occidental Petroleum's Role in Ongoing Crisis." Press Release. August 10, 1998, <http://www.hartford-hwp.com/archives/41/171.html>.

Spain, Peter. Personal interview at GreenCOM headquarters, Washington, D.C., October 20, 1999.

Speckhardt, Lisa. "Litigation—An Essential Tool for Environmental Protection." *EarthFocus: Friends of the Earth's News Magazine* 30, no. 1 (Spring 2000): 4–5.

Steinberg, Paul F. *Environmental Leadership in Developing Countries: Transnational Relations and Biodiversity Policy in Costa Rica and Bolivia.* Cambridge: The MIT Press, 2001.

Stephens, Joe, and David B. Ottaway. "How a Bid to Save a Species Came to Grief." *The Washington Post,* May 4, 2003, A-1.

———. "Nonprofit Sells Scenic Acreage to Allies at a Loss." *The Washington Post,* May 4, 2003, A-1.

Stoett, Peter J. "Global Environmental Security, Energy Resources and Planning: A Framework and Application." *Futures* 26, no. 7 (September 1994).

Studer, Marie. Personal interview with Chief Scientist, Earthwatch Institute headquarters, Maynard, Mass., August 30, 2002.

Swanson, Stevenson. "Countries Gather at United Nations for Earth Summit." *The Chicago Tribune,* June 26, 1997.

Swerdlow, Joel L. "Biodiversity: Taking Stock of Life." *National Geographic* 195, no. 2 (February 1999): 2–5.

Switzer, Jacqueline Vaughn, with Gary Bryner. *Environmental Politics: Domestic and Global Dimensions.* New York: St. Martin's Press, 1998.

Tamiotti, Ludivine, and Matthias Finger. "Environmental Organizations: Changing Roles and Functions in Global Politics." *Global Environmental Politics* 1, no. 1 (February 2001): 56–76.

Thomas, Caroline, ed. *Rio: Unraveling the Consequences.* London: Frank Cass, 1996.

Thoreau, Henry David. *Walden and Other Writings.* New York: Modern Library, 1950.

Thorne, Christopher G. *Border Crossings: Studies in International History.* New York: Basil Blackwell, 1988.

Tickner, Joel, Carolyn Raffensperger, and Nancy Myers. "The Precautionary Principle in Action: A Handbook First Edition." Science and Environmental Health Network, May 31, 2003, <http://www.biotech-info.net/precautionary.html>.

Time Europe Magazine. "Heroes 2003." *Time Europe,* April 28, 2003, <http://www.time.com/time/europe/hero/index.html>.

Tolba, Mostafa K., with Iwona Rummel-Bulska. *Global Environmental Diplomacy: Negotiating Environmental Agreements for the World, 1973–1992.* Cambridge: The MIT Press, 1998.

Truman, David. *The Governmental Process.* New York: Alfred A. Knopf, 1951.

Turner, John, and Jason Rylander. "Land Use: The Forgotten Agenda." In *Thinking Ecologically: The Next Generation of Environmental Policy.* Edited by Marian R. Chestow and Daniel C. Esty. New Haven: Yale University Press, 1997.

Turner, Tom. *Justice on Earth: Earthjustice and the People It Has Served.* White River Junction, Vt.: Chelsea Green Books, 2002.

Turner, Tom. Telephone Interview with Senior Editor, Earthjustice Legal Defense Fund, Oakland, Calif., June 2, 2003.

Tzoumis, Kelly. *Environmental Policymaking in Congress: The Role of Issue Defin-
itions in Wetlands, Great Lakes, and Wildlife Policies.* New York: Routledge,
2001.

Union of International Associations, ed. *Yearbook of International Organizations,
1998/99,* vol. 3. Brussels: K.G. Saur, 1999.

————. *Yearbook of International Organizations, 1996/1997.* Brussels: K.G. Saur,
1997.

United Nations (Press Release HE/916). "UNEP Releases First Global Biodiversity Assessment Report." November 14, 1995, available on WRI's Web
site February 22, 2000, <http://www.wri.org/biodiv/gba-unpr.html>.

United Nations Framework Convention on Climate Change. June 3, 2003,
<http://unfccc.int/>.

UN Security Council. "Note by the President of the Security Council." UN Doc.
No. S/23500, 1992, p. 3.

Uvin, Peter. "Scaling Up the Grassroots and Scaling Down the Summit: The Re-
lations Between Third World NGOs and the UN." In *NGOs, the UN, and Global
Governance.* Edited by Thomas G. Weiss and Leon Gordenker. Boulder: Lynne
Rienner Publishers, 1996.

Vernon, Raymond. *Sovereignty at Bay.* New York: Basic Books, 1971.

Verolme, Hans. Personal interview with Coordinator of Biodiversity Action Network (BIONET) at Washington, D.C., headquarters, July 28, 1999.

Victor, David G., Kal Raustiala, and Eugene B. Skoknikoff. *The Implementation
and Effectiveness of International Environmental Commitments: Theory and
Practice.* Cambridge: The MIT Press, 1998.

Voices of the Pantanal: The Story of the Hydrovia. Videocassette. Exec. Prod. & Dir.
Haroldo Castro. Prod. Flavia Castro. Nar. Dick Bertel. 16 min. Conservation
International, 1996.

Waltz, Kenneth. *Theory of International Politics.* Reading, Mass.: Addison-Wesley,
1979.

Wapner, Paul. *Environmental Activism and World Civic Politics.* Albany: State University of New York Press, 1996.

Web of Life: Exploring Biodiversity. Videocassettes (2). Exec. Prod. Gregory An-
dorfer. 120 min. Alexandria, Va.: PBS Adult Learning Satellite Service
(WQED/Pittsburgh in association with the World Wildlife Fund), 1995.

Weeks, W. William. *Beyond the Ark: Tools for an Ecosystem Approach to Conservation.* Washington, D.C.: Island Press, 1997.

Weiss, Thomas, and Chopra, Jarat. "Sovereignty Under Siege: From Intervention
to Humanitarian Space." In *Beyond Westphalia: State Sovereignty and International Intervention.* Edited by Gene M. Lyons and Michael Mastanduno. Baltimore: Johns Hopkins Press, 1995.

Wells, Donald T. *Environmental Policy: A Global Perspective for the Twenty-First
Century.* Upper Saddle River, N.J.: Prentice-Hall, Inc., 1996.

Wendt, Alexander. "Anarchy is What States Make of It: The Social Construction of
Power Politics." *International Organization* 46 (1992): 391–425.

Werbach, Adam. *Act Now, Apologize Later.* New York: HarperCollins, 1997.

Westman, Walter E. "Managing for Biodiversity: Unresolved Science and Policy Questions." *BioScience* 40, no. 1 (January 1990): 31.

White, Frank. *The Overview Effect.* Boston: Houghton Mifflin, 1987.

Wigley, Georgina, and Heather Baser. "The Notion of Effectiveness: Lessons from the Field of International Development." In *Protecting Biological Diversity: Roles and Responsibilities.* Edited by Catherine Potvin, Margaret Kraenzel, and Gilles Seutin. Montreal: McGill-Queen's University Press, 2001.

Wilder, Robert Jay. *Listening to the Sea: The Politics of Improving Environmental Protection.* Pittsburgh: University of Pittsburgh Press, 1998.

Wilson, Edward O. "The Column: Harvard University Press." *New York Times Book Review,* January 14, 1979, 43.

———. *Consilience: The Unity of Knowledge.* New York: Alfred A. Knopf, 1998.

———. "The Current State of Biological Diversity." In *Sources: Notable Selections in Environmental Studies.* Edited by Theodore D. Goldfarb. Guilford, Conn.: Dushkin Publishing Group, 1997.

———. *The Diversity of Life.* New York: W.W. Norton, 1992.

———. *In Search of Nature.* Washington, D.C.: Island Press, 1996.

———. *Naturalist.* Washington, D.C.: Island Press, 1994.

Wirth, Timothy. "Remarks at Conference on Nature and Human Society, National Academy of Sciences." Washington, D.C., October 30, 1997, transcript available, 5 pages, April 7, 1999, <www.state.gov/www/global/oes/971030tw .html>.

Wolin, Sheldon. "Editorial." *Democracy* 2 (July 1982): 2–4.

Wolk, Martin. "Starbucks Hops on Bandwagon with Eco-Friendly Coffee." *Planet Ark,* August 4, 1999, <http://www.planetark.org/dailynewsstory.cfm?newsid =2784&newsdate=04-Ag-1999.html>.

World Commission on Environment and Development. *Our Common Future.* Oxford: Oxford University Press, 1987.

World Resources Institute. "About WRI." April 7, 1999, <http://www.wri.org/ wri/wri.html>.

———. "Global Topics: Forests, Grasslands, and Drylands." June 2, 2003, <http:// www.wri.org/forests/key_issues.html>.

———. "WRI Article: Building Biodiversity Awareness in Schools." June 25, 2000, <http://www.wri.org/wri/biodiv/b33-gbs.html>.

World Wildlife Fund. "Member Tours at World Wildlife Fund-US." E-mail correspondence with author, July 13, 2000.

———. "World Wildlife Fund Action Center." June 15, 2000, <http://www .worldwildlife.org/actions/actioncenter.cfm>.

———. "World Wildlife Fund at Work." N. p., June 1999, 12, 13.

———. "World Wildlife Fund Travel." *Focus* 22, no. 2 (March/April 2000): 2, 7.

Worster, Donald. *The Wealth of Nature: Environmental History and the Ecological Imagination.* New York: Oxford University Press, 1993.

Wurman, Richard Saul. *Washington DC Access.* New York: Access Press, 1996.

WWF Conservation Action Network. "Mexico Cancels Salt Works Planned for Pristine Ecoregion of Importance to Gray Whales." March 15, 2000, results archive, <http://takeaction.worldwildlife.org/results/thanks.asp>.

WWF Global Network. "Dams are Direct Cause of Species Decline, says WWF." April 5, 2000, <www.panda.org/news/press/news.cfm?id=1910>.

Yanarella, Ernest J. "Environmental vs. Ecological Perspectives on Acid Rain: The American Environmental Movement and the West German Green Party." In *The Acid Rain Debate: Scientific, Economic, and Political Dimensions*. Edited by Ernest J. Yanarella and Randall H. Ihara. Boulder: Westview Press, 1985.

Yanarella, Ernest, and Richard S. Levine. "Does Sustainable Development Lead to Sustainability?" *Futures* 18 (October 1992): 759–74.

Yeager, Brooks. Personal interview with Vice President, Global Threats Program, World Wildlife Fund headquarters, Washington, D.C., October 17, 2002.

Yellowstone National Park. "Yellowstone in the Afterglow." June 3, 2003, <http://www.nps.gov/yell/publications/pdfs/fire/htmls/chapter1.htm>.

Yin, Robert K. *Case Study Research: Design and Methods*. Newbury Park, Calif.: Sage Publications, 1989.

Young, Nina. "Statement by Nina Young, Marine Mammalogist, Center for Marine Conservation." February 3, 2000, <http://www.environmentaldefense.org/pubs/Filings/96.04.30a_tuna.dolphin.htm>.

Young, Nina M., W. Robert Irvin, and Meredith L. McLean. "The Flipper Phenomenon: Perspectives on the Panama Declaration and the 'Dolphin Safe' Label." *Ocean and Coastal Law Journal* 3 (1997): 57–115.

Young, Oran R., ed. *Global Governance: Drawing Insights from the Environmental Experience*. Cambridge: The MIT Press, 1997.

———. *International Cooperation: Building Regimes for Natural Resources and the Environment*. Ithaca: Cornell University Press, 1989.

———. "Regime Effectiveness: Taking Stock." In *The Effectiveness of International Environmental Regimes: Causal Connections and Behavioral Mechanisms*. Edited by Oran R. Young, Cambridge: The MIT Press, 1999.

Young, Oran R. And Marc A. Levy. "The Effectiveness of International Environmental Regimes." In *The Effectiveness of International Environmental Regimes: Casual Connections and Behavioral Mechanisms*. Edited by Oran R. Young, Cambridge: The MIT Press, 1999: 1–32.

Zacher, Mark W. "The Decaying Pillars of the Westphalian Temple: Implications for International Order and Governance." In *Governance without Government: Order and Change in World Politics*. Edited by J. Rosenau and E. O. Czempiel, 58–101. Cambridge: Cambridge University Press, 1992.

Index

Numbers in *italics* refer to illustrations. Page references to tables are followed by *t*.

Montiel, Rodolfo, 163
Montreal Protocol, 29
Moore, Franklin, 78, 161, 184
Moore, Gordon, 143
Morgenthau, Hans, 199n
Moulaert, Azur, 103
mountain effect, 109–10
Muffett, Carroll, 41, 168
multinational corporations, 78–79

National Audubon Society, 27
National Environmental Policy Act
    (NEPA; 1969), 58
National Geographic, 160–61
National Invasive Species Council, 169
National Marine Debris Monitoring
    Program (NMDMP), 86–87
National Resources Defense Council
    (NRDC), 28, 29
National Wildlife Federation, 27
natural disasters, 11–12
Nature Conservancy, The (TNC).
    See The Nature Conservancy
Nepal, 123–24
networking: among NGOs, 145–47.
    See also grassroots networking
Newton, Sir Isaac, xvii
NIMBY (not-in-my-backyard)
    attitudes, 98
Noah (biblical), 183
North American Free Trade Agree-
    ment (NAFTA; 1994), 10, 118,
    197n
Northern Hemisphere: discrepancies
    in biodiversity in Southern vs., 9;
    NGOs in, 4
Nye, Joseph, 10, 25–26

Occidential Petroleum, 60
Ocean Conservancy, The: awards given
    to, 153–54; on dolphin-safe tuna,
    59; domestic/international linkages
    by, 68–70; Earthjustice Legal De-
    fense Fund and, 156; International
    Coastal Cleanup program of, *71,*
    122, 133, 212n; Internet used by,

103; name of, 206n; short-term/
    long-term linkages by, 86–87; staff
    turnover in, 165
Olson, Mancur, 149, 182, 217n
Ophuls, William, 17
Oppenheimer, Michael, 153
Orr, David, 184
Ottaway, David, 79
overhead expenses for NGOs, 169–72,
    171*t*

Paarlberg, Robert, 204n
Panjabi, Ranee, 30
Parks in Peril Program (PIP), 66–68
Parque Nacional del Este (Dominican
    Republic), 67, *67*
participatory democracy, 183–86
participatory strategies, 47, 97–99,
    99*t;* contributions to linkages of,
    135*t;* domestic/international link-
    ages in, 117–24; effectiveness of,
    in establishing linkages, 136*t;*
    grassroots networking, 99–100,
    106–9; Internet used for, 100–
    104; mainstream strategies used
    with, 137–38; mainstream strate-
    gies versus, 72; organizational
    constraints on, 169–75; organiza-
    tional supports for, 159–65; Sierra
    Club's Grand Canyon campaign,
    95–97, *96;* strengths and weak-
    nesses of, 116*t*
partnerships among NGOs, 145–47,
    151*t,* 154–65, 168
Paul, Seema, 143, 146
Peace of Westphalia (1648), 22, 200n
pharmaceutical industry, 38, 125
Plotkin, Mark, 216–27n
Plummer, Mark, 13, 14
political issues: lobbying, 56–58; sov-
    ereignty and regimes, 21–24, 179;
    in species loss, 20
pollution: marine pollution, 87
Porter, Gareth, 27, 29, 30
Prakash, Reshma, 68–69
precautionary principle, 7

MICHAEL JR., holds a Ph.D.
in Political S aduate certificate in
Environmen m the University of
Kentucky. H bachelor's degree
from Vander An outdoor enthu-
siast and for it, Gunter is currently an Assista Rollins College in
Winter Park,